SEPTIC TANK PRACTICES

Peter Warshall

With the generous assistance
of J. T. Winneberger & Greg Hewlett

SEPTIC
TANK
PRACTICES

ANCHOR BOOKS

Anchor Press/Doubleday, Garden City, New York, 1979

Septic Tank Practices was originally published by the Mesa Press in 1973. The Anchor Press edition is revised and expanded.

Anchor Press edition: 1979

Library of Congress Cataloging in Publication Data
Warshall, Peter.
  Septic tank practices.

  Bibliography: p. 167.
  1. Septic tanks. I. Title.
TD778.W37   628'.742
ISBN: 0-385-12764-2
Library of Congress Catalog Card Number 77–76288

For my town's people who kindly and humorously let me snoop around their backyards, check their water use, peek and plunge into their septic tanks, and learn with them the mysteries of soils, drainage, and water purification.

CONTENTS

Sewer manhole cover.

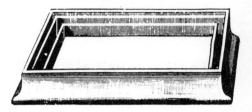

SEPTIC TANK PRACTICES

1865

1870

Just last year Prince Philip walked into the ladies' room of a British railway station. Surprised, he found no signs saying "Ladies" or "Gentlemen." He found out that Queen Victoria was embarrassed by her subjects going to the bathroom. She ordered all signs removed from her presence. So every time the royal train arrived at a station, all signs were covered or removed. Prince Philip has finally rescinded the decree.

On the farm of Mrs. Wu, near Kashing, we were surprised to observe that one of the duties of the lad who had charge of the cows was to use a six-quart wooden dipper with a bamboo handle six feet long to collect all excreta, before they fell upon the ground, and transfer them to a receptacle provided for the purpose. There came a flash of resentment that such a task was set for the lad, for we were only beginning to realize to what lengths the practice of economy may go, but there was nothing irksome suggested in the boy's face. He performed the duty as a matter of course and as we thought it through there was no reason why it should have been otherwise. In fact, the only right course was being taken. Conditions would have been worse if the collection had not been made. It made possible more rice. Character of substantial quality was building in the lad which meant thrift in the growing man and continued life for the nation.

—*From* Farmers of Forty Centuries—*1908 (see Bibliography)*

When hygiene was commercialized, the last hope that sewage could be reused and handled sensibly was lost. (From The Underground Sketchbook *of Tomi Ungerer,* Dover paperback, 1973, a wonderful series of taboo-breaking cartoons.)*

PROLOG

Sanitation in the Middle Ages, from an old woodcut.

As industrialization intensified, Western civilization became more and more alienated from the body's plumbing and its connection to Nature's pathways. Instead of eating, defecating onto the ground, fertilizing plants with feces, and eating again, we simply reach behind our backs and pull a little chromium lever. Instead of defecating into the earth, we sit on a toilet filled with good drinking water which comes from some unknown river and, after flushing, goes to some unknown destination. Instead of taking responsibility for our excrement, we are embarrassed by defecation and avoid direct discussion by substituting all kinds of diversionary vocabulary ("caa-caa" and "poo-poo" or abbreviations "Number 1" and "Number 2" or sidewise expressions like "May I be excused?").

Western civilization is caught in a paradox. While it has the most modern sanitary equipment, its mentality has become more primitive than any preliterate society. While each citizen consumes more food and water than any previous society, these

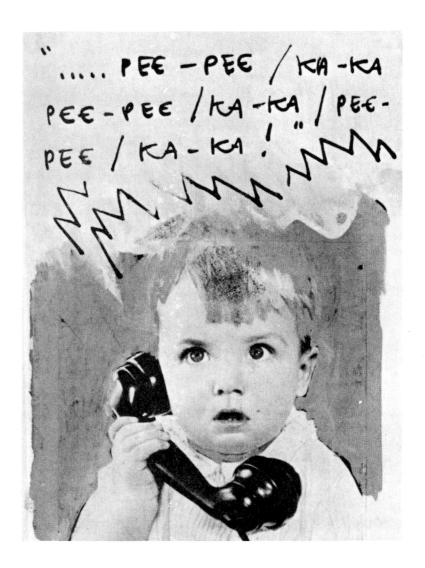

Collage by Joe Brainard

same citizens are totally ignorant of the simplest and most basic concerns of humanity since before the Pleistocene: Where does your water come from? Where does it go after you urinate? Where does your food come from? Where does it go after you defecate? Whose life is changed and how by your feces' destination? Compared to the people of the Dark Ages of Europe, today's New Yorker or Los Angeles resident has the more "primitive" mind. Turn the faucet, and magically "clean water" appears. Flush, and "magically" everything empties into the unknown.

Four events have created our alienation from the circle of feces-fertilizer-food-feces. Piped water eliminated the outhouse and chamber pot and led to the invention of the flush toilet. The flush toilet itself allowed the Victorian flush-and-forget mentality to flourish. Modern medicine, having conquered all the dangerous diseases caused by sewage, reduced the necessity to connect feces with the human body. Finally, the use of petroleum-based fertilizers temporarily broke the necessity of using feces for growing food.

By the mid-forties, excrement became known as "waste" and flushing was "waste disposal"—all part of the same semi-embarrassed (though now couched in technical-sounding lingo) vocabulary. By the fifties America had developed a full-blown "excrement taboo" which was widely commercialized and exploited by the cosmetic industry. Adding sex to embarrassment, any excrement (sweat, exhalation, even tears) was to be hidden, disguised, and buried in pine scent.

Meanwhile, the people of Japan, Korea, and China regarded feces as "night soil"—as fertilizer to be given freely back to the land. In prewar Japan, rent was lowered in Hiroshima in proportion to the amount the house privy was used. In Vietnam, outhouses are still placed along the road with signs encouraging travelers to use the facilities. A balance existed between city "wastes" which were transported to farms for fertilizer, and the food, fertilized by human manure, returned to the cities.

The era of flush-and-forget is ending. New Yorkers cannot avoid the sewage sludge returning to their shores. Midwesterners cannot but regret the loss of fishing and swimming in polluted

Lake Michigan. Californians are increasingly aware that sea urchins—fed by a continuous supply of rich city sewage—are devastating offshore kelp, ruining both the nursery of many fish and a profitable industry. The United States has become the world leader in dumping rich sewage into rivers, lakes, and oceans while digging up the rest of the world for petroleum-based fertilizers.

We are learning: wastes are not wastes. They are misplaced natural resources. We cannot "dispose" of anything on Earth except by sending it to outer space. "Wastes" can only re-enter the Nutrient and Water Cycles on this planet. We cannot avoid our "wastes." Because we exist in the biosphere of Earth and are connected to the Nutrient and Water Cycles of the planet, "wastes" return—usually with a vengeance. But as humans we can encourage "wastes" and "wastewater" to re-enter only certain, specific natural cycles and communities where they can benefit us and other living creatures and plants. By understanding our wastes, by connecting our body's plumbing to Nature's pathways, we can eliminate current practices that damage both our bodies and planet life.

This book is dedicated to a rebirth of responsibility toward earth, air, and water; to breaking the excrement taboo; and to healing the somewhat disconnected and schizophrenic mentality that hinders this reawakening. If you eat, you defecate. And it is the responsibility of every citizen to make sure that defecation means fertilization of the land that feeds us.

> *When we reflect upon the depleted fertility of our own older farmlands, comparatively few of which have seen a century's service, and upon the enormous quantity of mineral fertilizers which are being applied annually to them in order to secure paying yields, it becomes evident that the time is here when profound consideration should be given to the practices the Mongolian race has maintained through many centuries, which permit it to be said of China that one-sixth acre of good land is ample for the maintenance of one person, and which are feeding an average of three people per acre of farmland in the three southernmost of the four main islands of Japan.*

> —*From* Farmers of Forty Centuries—*1908.*

INTRODUCTION

The Good Daughter-in-Law by Hsieh Chang-yi

Early in the morning, the magpies cry,
The newly-wed daughter-in-law is carrying excreta on a pole
Liquid from the excreta stains her new trousers
The hot sweat soaks into her embroidered jacket
The commune members praise her and mother is pleased
All tell her she has got a good daughter-in-law.

—Folk song translated by J. Gittings

In the past two decades there has been a bandwagon of pressures pushing small communities to sewer up. Health officials said that home-site systems such as septic tanks were just unmanageable, and sure-bet health hazards. Engineering firms peddled big sewers to make profits that were impossible with *at-home* sewage treatment. Water companies favored big sewers because they used more water. Real estate agents wanted big sewers because they allowed smaller lots. The federal and state governments liked them because providing more big sewers was equated with progress and a higher standard of living. Even the Environmental Protection Agency, until recently, believed centralized sewers would lead to less pollution than home-site sewage treatment. Between 1950 and 1970, ten million homes with home-site sewage treatment were connected to new centralized sewer systems.

This is perhaps the first book to argue that home-site sewage treatment pollutes less, costs less, uses up less of our energy resources, and is less of a health hazard than centralized sewage treatment. In addition, home-site sewage systems treat domestic sewage in soils, and soils are by far the best purifiers known to man. We hope to persuade health officers, government agencies, and communities that home-site sewage treatment is superior to centralized sewerage (chapters 1 and 10).

At the same time, and equally important, this book tries to make home-site sewage treatment as ecologically and technically sound as possible. Every aspect of septic tank and drainfield design from materials to maintenance is explained, using information from the last twenty years of research (chapters 5 through 9).

———————————— NUTRIENT FLOW: OLD ————————————

An average U.S. citizen contributes 3.5 pounds of phosphorus and 9.9 pounds of nitrogen to water each year. In addition, each citizen is responsible for his/her share of agricultural and livestock "wastes" produced during the production of food. In total, each citizen discharges 4.4 pounds of phosphorus and 26 pounds of nitrogen into our oceans, lakes, and rivers. This is enough to fertilize one ton of living plants. A city of one million discharges enough nutrients to create one million tons of living plants.

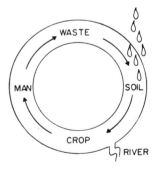

We will necessarily disagree with the Public Health Service *Manual of Septic Tank Practice* because, frankly, this publication is now obsolete. The PHS has not updated their information and has been sadly lax in spreading new information among the people of this country.

We will not *push* septic tanks as the best and only home-site sewage treatment. Chapter 2 clearly indicates that other kinds of home-site systems are preferable to septic tanks in many circumstances. Throughout, there are warnings about septic tank/drainfield limitations (especially in chapters 6 and 7), and alternatives like evapotranspiration beds are suggested.

Finally, the soil community is the key to clean water. All of chapter 3 is dedicated to understanding how soils clean wastewater and help recycle both nutrients and water. This chapter connects Reason to Nature and provides the groundedness necessary for good design, installation, care, and maintenance of home-site sewage treatment.

NUTRIENT FLOW: NEW

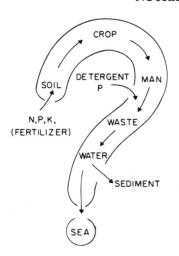

The misuse of nutrients comes from bypassing soil, the introduction of automatic washing machines with phosphate-based detergents, and the use of synthetic fertilizers to replace human and animal manures. The passage of nutrients into lakes and oceans has upset natural balances causing fish kills in Lake Erie and the ruination of the kelp industry in parts of California.

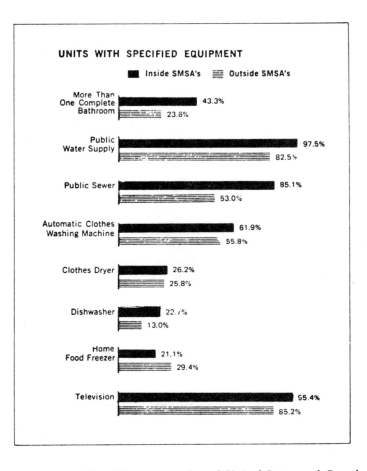

UNITS WITH SPECIFIED EQUIPMENT

■ Inside SMSA's ☰ Outside SMSA's

More Than One Complete Bathroom
43.3%
23.6%

Public Water Supply
97.5%
82.5%

Public Sewer
85.1%
53.0%

Automatic Clothes Washing Machine
61.9%
55.8%

Clothes Dryer
26.2%
25.8%

Dishwasher
22.7%
13.0%

Home Food Freezer
21.1%
29.4%

Television
95.4%
85.2%

About 50 per cent of rural United States and Canada and 15 per cent of urban America use home-site sewage systems. This is one-quarter of America's housing (20 million dwelling units) in which about 50 million Americans live. Until the oil shortages, inflation, and recession, about 300,000 new homes with home-site sewage treatment were being built each year. The black bars show metropolitan percentages. The striped bars show rural percentages. Data from U. S. Dept. of Commerce, 1970 Census. Note bias of the questionnaire. The Bureau of Census automatically regards "public power" (vs. septic tank) as a desirable.

1. SEWAGE AND SEWAGE TREATMENT

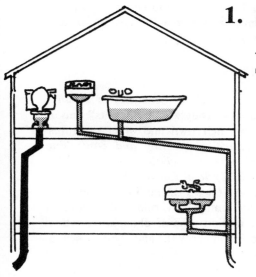

A house with grey water separated from black water. This kind of plumbing makes sewage treatment easier and water recycling easier. Chapter 4 gives the details.

Household sewage is simply everything a family flushes down the toilet and washes down the drains. About 60% of sewage wastewater comes from lightly polluted sources. This water, called *grey water,* drains from the laundry, kitchen, and bathroom sink, shower, and/or bathtub. The other 40% of sewage wastewater comes from the toilet. Since this water carries feces out of the house, it is considered heavily polluted. It is called *black water.*

Sewage is 99.9% water. In other words, sewage is highly diluted because we use so much water (50 to 350 gallons per person per day) just to flush and wash away a tiny amount of "undesirable" solids (16 to 100 ounces per person per day).

One tenth of one per cent of sewage is solid substance. This one tenth of one per cent includes both organic (80 to 85%) and inorganic (15 to 20%) substances. Organic substances come from feces and urine, detergents, soaps, and food wastes (especially in homes with garbage grinders). The inorganic substances originate from household chemicals: water softeners, borax and chlorine, paints, and photo chemicals.

What Is a Pollutant?

These substances in sewage can be either beneficial or harmful to human and other living creatures. They are "pollutants" only if humans *cannot* recycle them for their own and other creatures' benefit. *Sewage treatment is basically directing and controlling the recycling of excrement and other household substances back into the Water Cycle and Nutrient Cycle of Earth.*

For instance, if all of North America's 22 million on-site sewage treatment systems were functioning well, one million tons of fixed nitrogen and one-quarter million tons of phosphorus would recycle each year for better plant growth and better evapotranspiration. Recycling is easier with home-site sewage because it occurs *in soil*. This recycling would save about 8 to 9 million tons of petroleum-based commercial fertilizer.

On the other hand, centralized sewers waste a minimum of 1.1 million tons of fixed nitrogen because sewage is released *in surface waters* (lakes, rivers, oceans). This wasted nitrogen is equivalent to about 10 million tons of petroleum-based commercial fertilizer.

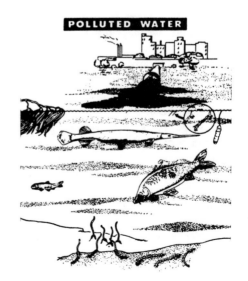

POLLUTED WATER

By not recycling quickly in soil, the nitrogen upsets natural communities, causes unwanted and suffocating algae growths, can ruin various economic enterprises like the kelp industry, and can even cause a rare disease in infants called methemoglobinemia. (This disease produces "blue babies" and can be passed from mother to infant during breast feeding.) Nitrogen, in home-site sewage treatment, is a beneficial resource; in centralized sewers, it produces a horror show. (More in chapter 10.)

Biological Oxygen Demand

All the organic and many of the inorganic substances in sewage are large, complicated molecules. They must be broken down by sewage treatment. This process requires oxygen. The oxygen required to treat sewage is a "demand" on the dissolved oxygen in the water. If the demand for oxygen is too great, the sewage goes untreated and polluted water will enter lakes and oceans. This deprives animals such as fish and shrimp of their oxygen. In extreme situations, like when a pipe from a city sewer runs directly into the ocean, fish and other animals suffocate. Natural wildlife is greatly reduced and many natural balances are totally upset. This demand for oxygen to break down large molecules is one of the main concerns of sewage treatment. Not surprisingly, it is called the *biological oxygen demand,* or *BOD.*

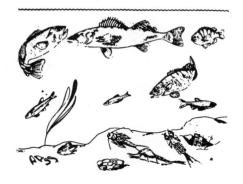

Pathogens

Finally, microbes are found in sewage. Bacteria, viruses, and fungi—some dead, some alive—come from your stomach, blood, lungs, and even your skin. A small number of these microbes cause human disease. These are called *pathogens*.

Sewage Treatment

To recycle sewage for the benefit of all living creatures, sewage must be altered with human help. Humans must:

1. Reduce the amount of sewage by reducing water use, which in turn will reduce the difficulties and costs of sewage treatment.
2. Help break down the solids so that microbes can more easily digest them and release the nutrients quickly into the natural cycles.
3. Kill the disease-causing microbes so that other humans will not be harmed.
4. Use as much of the organic substances as possible before they travel by soil or water to some natural community where their presence upsets the existing balance of plant and animal (including human) life (i.e., reduce BOD).
5. Remove harmful chemicals (like DDT) before they harm plants and animals (including us) by accumulating in the food chains.

All domestic sewage treatment systems should be judged by these standards and the ability to attain these standards cheaply.

FIG. 220. — INTERIOR OF A PRIVY IN ASAKUSA.

2. HOME-SITE SEWAGE TREATMENT

Spaceship technology has brought to earth an incredible variety of appliances and contraptions for sewage treatment. There are destrolets that burn excrement to ashes. These can be attached to motor homes, and the fumes, filtered through the exhaust pipes, meet air pollution standards. There are freeze toilets that freeze excrement into ice-blocks that can be carted away by trucks. There is even a Swedish "sausage toilet," which catches excrement in plastic baggies that come on a roll, then seals them for the garbage collector. We are not concerned with these high-tech solutions because they are very costly, require extravagant amount of

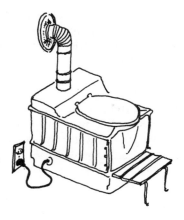

An electric compost privy by Mulbank in Sweden. Costs over $750 plus 10¢ each day for electricity. Must be composted or mixed in garden soil after treatment.

electricity or petroleum-based fuels, never recycle, and, with so many parts, tend to break down. Instead, we will treat four low-tech, small-scale sewage treatment systems. All treat sewage at the home site, each in its own way. They are the pit privy (outhouse), the compost privy, the septic-tank/drainfield system, and the aerobic home unit.

Pit Privy

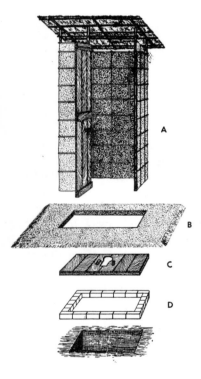

The *pit privy,* or outhouse, uses no water. The absence of water greatly reduces the volume of sewage and pollution potential, and simplifies treatment and recycling. The pile of excrement decomposes: faster when it's warm; slower when it's cold. Decomposition reduces bulk, so the privy can be used for a long time. Extremes of seasonal heat or cold kills some disease-causing microbes in the decomposing pile of feces. But, even more important, the time spent without us, the protective human host, means most pathogens die. In addition, as the pile decomposes, some of the microbes and nutrients find their way into the surrounding soil. The soil, the planet's most astounding filter, absorbs and strains out the nutrients and remaining disease-causing organisms (see chapter 3). A privy that is well placed (away from water) and fly-proofed cannot pollute or be unsanitary.

In summary, the pit privy is the most reliable, most easily managed and replaced, cheapest, and least polluting of all home sewage treatments. Where water is scarce or the water system is not pressurized, the pit privy is the only device that makes sense.

Compost Privy

The *compost privy* also decomposes excrement *and* all other organic household wastes, such as vegetable scraps. Like the pit privy, it does *not* use water. But the process of decomposition in a compost privy is very different from that in a pit privy. The pile is not left to rest. It is aerated either by turning it with a shovel or by forcing air through perforated pipes in the pile. The extra air (and, in some cases, extra heat) means the decomposition is faster and produces fewer odors than the pit privy. The composted wastes are not allowed to filter into the soil. Instead, after a year they are shoveled out of the compost privy and used directly for fertilizer. No pathogens survive a year of composting.

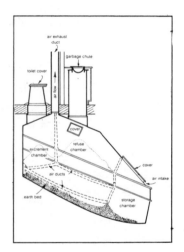 

The Clivus Composting Toilet (left) and the Toa-Throne (right). Both are totally enclosed, fiber-glass privies. They digest feces, urine and kitchen scraps aerobically through ventilating pipes. They are officially acceptable in Maine and experimentally acceptable in five or six other states. Costs vary from $750 to over $2,000. Excellent privies for city use. Information on Clivus from 14A Eliot Street, Cambridge MA 02138 and Toa-Throne from Enviroscope, Inc., P.O. Box 752, Corona del Mar, CA 92625.

Compost privies are reliable but need lots of attention: keeping the right mixture of feces with other organic wastes, as well as insuring proper aeration and warmth. There is absolutely no pollution possible, because compost privies are self-contained units. There are both expensive and cheap compost privies available. Problems have occurred in these enclosed privies with too much urine causing pounding and odors as well as flies from loose seals and vegetable scraps. Pathogen survival and dose levels need more research. So far, no health hazard has been demonstrated.

An additional expense for both pit and compost privies may be a greywater system. To repeat, greywater is lightly polluted because it has no feces. It is the wastewater from sinks, baths, and washers. Some kind of disposal system is needed for this water— usually just a pit. But some health departments require expensive greywater systems (see chapter 3). Some health departments also ban compost privies.

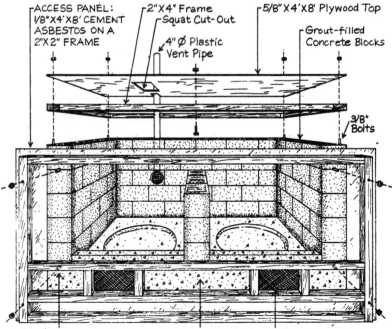

ACCESS PANEL: 1/8"X4'X8' CEMENT ASBESTOS ON A 2"X2" FRAME — 2"X4" Frame — Squat Cut-Out — 4" ∅ Plastic Vent Pipe — 5/8" X4'X8' Plywood Top — Grout-filled Concrete Blocks — 3/8" Bolts

3/8" Bolts set in mortar joint   8"X4'X8' Concrete Slab   6"X12" Screened Vent

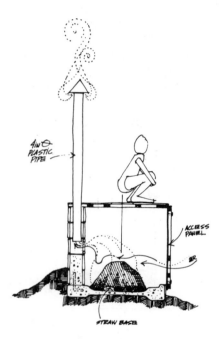

First chamber is used for 6 months.

The *Farallones Institute Composting Privy (Tech Bulletin No. 1, Farallones Institute, 15290 Coleman Valley Road, Occidental, CA 95465, $3.00) is for rural homes. It needs a 4'×8' area. You build it and maintain it by following the procedure illustrated. It requires a greywater system as well as health-department approval. In use in California. Toilet or seat can be used with this privy if you are adamant against squatting. Maximum use is about 15 persons daily.*

After six months, the pile is moved to the aging compartment.

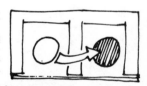

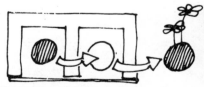

After one year, the first pile is ready for garden use.

Drum Privy

Since the commercial compost privies were so expensive, various Californians have developed a cheaper version. It consists of a coated 55-gallon drum. Usually, it is jacked up under the house. A plastic aerator reduces moisture. A vent pipe lets moisture and odors out into the atmosphere. It is alternated with a second drum. The filled drum can be composted in an already hot pile, buried, removed to a treatment plant or let to "cold compost" (moulder). The mouldering process requires at least one year.

Guidelines available from: DRUM PRIVY, Box 15, Overlook, Bolinas, Ca 94924 ($1.00) or Rural Wastewater Alternatives (see Bibliography).

CROSS SECTION
MATSON DRUM PRIVY

6" TO 12" OF SOIL OR SAWDUST TO FORM OUTER LAYER

MANURE

PLANT MATERIAL

SLOPE TO DRAIN OFF LIQUIDS

CENTER OF PILE MATERIAL FROM DRUM COMPOSTER

MANURE

PLANT MATERIAL

MANURE

ALTERNATE LAYERS OF ANIMAL MANURE & PLANT MATERIALS

GROUND

PLANT MATERIAL– INITIAL LAYER

AIR-PERMEABLE BASE (SOIL, STRAW, SMALL ROCKS, ETC.)

Cesspool

The *cesspool* is essentially a kind of combined septic tank/drain-field. The hole, which now serves as both septic tank and drain-field, must be much larger than the hole needed for a septic tank and a drainfield separately. Cesspools tend to clog more quickly than septic tanks because solids fill up the soil pores. The basic principles of septic-tank/drainfield systems apply also to cess-pools, so no further special mention will be made of cesspools in this book.

Septic-Tank/Drainfield System

The *septic-tank system* is a home-site sewage treatment system that uses the flush toilet. Wastewater from all appliances (sinks, toilets, showers, etc.) is combined. Septic-tank systems can pollute more than compost or pit privies do because feces are diluted in water and the volume of waste is greatly increased.

The septic-tank system actually has two distinct sections: the septic tank itself and the drainfield. The tank is a box that eliminates at least half the excrement by allowing time for solids to settle and be eaten by microbes. The treatment occurs without much oxygen. It is slow but reliable. The wastewater (minus much of the solid matter) then passes into a hole in the ground. The hole can be of almost any shape and depth. The most common shape is a linear trench usually between three and six feet deep. This trench design is called the drainfield (or leachfield, filterfield, absorption bed, disposal or subirrigation field). The hole may also be pit-shaped or within a man-made mound.

The wastewater from the septic tank receives further treatment in the drainfield. The soil adsorbs viruses, strains out bacteria, filters large wastes, and chemically renovates them into nutrients that can be used by plants. In the drainfield, there is usually more oxygen for sewage treatment, as air circulates in and out of the soil. Treatment is reliable for the lifespan of the drainfield.

A *MOUND is a man-made drainfield when soils are poor or the water table is too high. The trench is built above the ground level by trucking in enough good soil to treat the septic tank's effluent. More in chapter 7.*

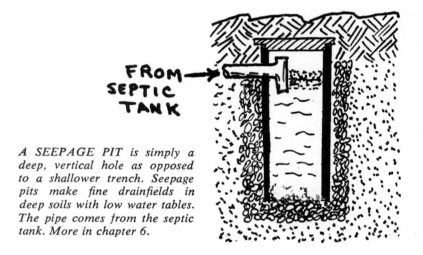

FROM → SEPTIC TANK

A *SEEPAGE PIT is simply a deep, vertical hole as opposed to a shallower trench. Seepage pits make fine drainfields in deep soils with low water tables. The pipe comes from the septic tank. More in chapter 6.*

The cleaned water from the drainfield is disposed of when it moves through the soil to aquifers and streams, evaporates from the ground surface, and is ingested by thirsty plants.

In summary, the septic-tank system requires a high initial cost for materials and installation compared to a pit privy. Its ability to pollute or cause unhealthy conditions is almost as low as the pit privy's. Septic-tank systems require a pressurized water supply, some maintenance, and a home surrounded by a garden, lawn, or other green space. Through subirrigation in the drainfield, septic-tank systems recycle the nutrients and the cleaned water.

Automated Sewage Treatment System

The Multi-Flo Waste Treatment System is typical of high-tech home devices. It has a pump, aerator, filter, and an additional charcoal filter. The cost is close to $3,000, plus all the yearly maintenance, and replacement expenses. Effluent still requires a drainfield, unless expensive chlorinators are added.

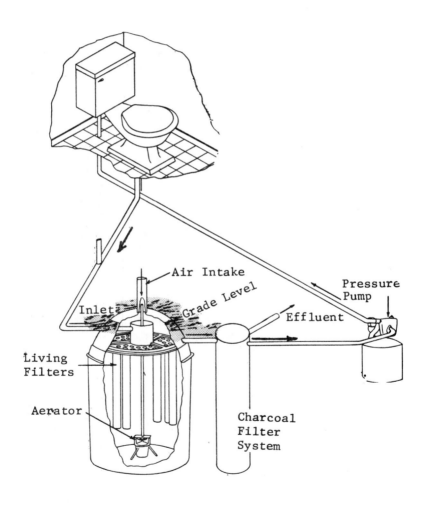

Home Aerobic Unit

The *home aerobic unit* is a recent invention that pumps air into the wastewater so that the biological oxygen demand will not become a problem. When working, home aerobic units also treat sewage faster and better than septic tanks. But these modern contraptions should be avoided. They are expensive, require continuous energy for the air compressor, can easily break down because of all the mechanical parts, and require high-cost maintenance. They all need cleaning once a year. In addition, they give unreliable treatment. If you have a party, the aerobic home unit can temporarily give up from the shock and discharge untreated sewage. Some manufacturers have tried to hold down the cost by saying that the sewage treatment is so good you can discharge directly into streams or on land. But most government authorities recognize the shock-loading problem and require a drainfield or chlorination—adding costs. Finally, while septic tanks remove 20 to 40 per cent of the nitrogen in sewage, aerobic units do not remove nitrogen or phosphorus or potassium to any significant degree. Discharging these nutrients directly into a stream is pollution.

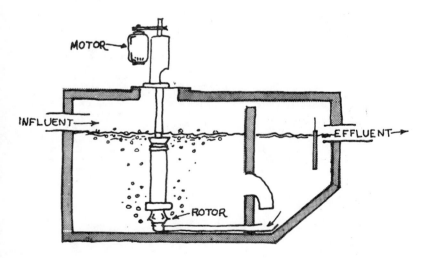

REL IAB ILI TY	PIT PRIVY	COMPOST PRIVY	SEPTIC TANK/ DRAINFIELD	AEROBIC UNIT
	Aerobic & anaerobic composting with some infiltration. Very stable.	Most aerobic composting. Stable when proper carbon/ nitrogen balance maintained.	Settling, flotation, & anaerobic digestion in tank. Aerobic & anaerobic filtration & digestion in drain-field. Very stable if not overloaded.	Aerobic digestion least stable due to "shock" loading and mechanical complexity.
MAINTEN- ANCE	Very easy. Minimal labor.	Not so easy. Proper amounts of vegetable matter must be added to feces. Manual labor required weekly in some models.	Easy. Labor minimal. Needs checking for pumping about every 2–4 years. If dual-field, needs yearly manual switching.	Difficult. Many mechanical parts needing specialist labor. Outside energy source can be a problem. Needs cleaning each year.
COSTS	INITIAL: Very inexpensive ($50). Greywater system may be required. OPERATION: None MAINTENANCE: None	INITIAL: See illustrations. Greywater system may be required. OPERATION: None MAINTENANCE: None	INITIAL: Pretty expensive ($800–$4,500) depending on size, contractor, and materials. Home-made can be cheaper. Greywater system used to advantage. OPERATION: Water costs. MAINTENANCE: Pumping every 3–10 years ($40–$85).	INITIAL: Very expensive ($1,600–$3,000+; drainfield, & filtration & chlorination not included). Greywater system used to advantage. OPERATION: Water costs. Electricity costs ($150+ each year). Filtration & chlorination ($300+ each year).
LIFESPAN	About 10 years for a family of 4.	As long as materials last (Farallones, 20 years; Clivus, 60 years?; drum, 10–15 years).	10–75 years depending on soils and design.	Less than 10 years before major parts replacement necessary.

TREATMENT: SUMMARY

	PIT PRIVY	COMPOST PRIVY	SEPTIC TANK/ DRAINFIELD	AEROBIC UNIT
POLLUTION AND SANITATION	If not near or in water, no pollution or sanitary problems.	No pollution. No sanitation problem with proper composting. Overloading may be a problem when daily use exceeds 4–6 people (55-gal. drum), 8–12 people (Clivus), 15 people (Farallones).	Larger pollution & sanitation problems because of water-feces mix. Soils, groundwater, slopes and overload can be problems outside property line because of in-soil discharge.	Larger pollution & sanitation problems because of water-feces mix. "Shock" loadings, mechanical breakdown, power blackout, & inadequate treatment can be problems. Above-soil discharge has more dangers than in-soil. In-soil has negligible pollution potential outside property line.
RECYCL-ING	Ultimately by burial & planting a tree. Adds nutrients & humus.	Great fertilizer for gardens, etc.	Subirrigation in drainfield ultimately fertilizes plants & may recharge water supplies.	ABOVE GROUND: water and nutrients recycled by irrigation. IN GROUND: recycled by absorption. In water courses, nutrients can be pollutants.
COMM-UNITY	Accommodates high densities in cities–if away from water. Politically, not available for city use. Discouraged even in rural areas.	Can accommodate high densities in cities. Need pick-up of compost. Politically, new to city & rural health depts. Rarely accepted with ease.	Low densities only—need green space for drain-field. Rural use. Accepted home-site system but many badly designed.	Low densities with in-soil discharge. High-density with above-soil discharge. Rarely accepted by health depts. because of erratic behavior.

Summary

If water is scarce, use a pit privy or compost privy. If water and good soil are available, it is still wise to use a pit or compost privy, because they are cheaper and save water. If you must combine greywater with feces or must have a flush toilet or want subirrigation through a drainfield, then consider the septic-tank system. Avoid home-site sewage treatment systems that have a continual need for outside energy and that have many moving, mechanical parts.

3. GOOD SOIL, CLEAN WATER

"*A teaspoon of living earth contains five million bacteria, twenty million fungi, one million protozoa, and two hundred thousand algae. No living human can predict what vital miracles are locked in this lot of life, this stupendous reservoir of genetic materials that has evolved continuously since the dawn of life on earth (about two billion years ago).*" (From *Clean Water*, by Leonard Stevens.)

Soil is the key to clean water. It is a living filter better than any filter man has invented. No material cleans nutrients or disease-causing microbes from wastewater as well as earth does. Soil works as a *physical* strainer, a *chemical* renovator, and a *biological* recycler of all wastewater passing through it. The use of soil is the oldest and most tried form of wastewater treatment—refilling lakes and underground aquifers with drinkable water as well as adding humus and food for the plant world. This chapter is a thumbnail sketch of the surface layer of the Earth's crust and how it does the job of cleaning and recycling household water.

Chapter drawing above shows a stalked and free-swimming ciliate (a kind of protozoon) in a soil of silt and sand. Bacteria and viruses can also be seen swimming in the space between the soil grains. The particles are about 100 microns wide; the viruses about 0.02 microns; the bacteria about 5 microns.

The Soil Community

Soil is not just an inert arrangement of pore openings among particles of dirt. Billions of microscopic creatures live among the textures and layers of soil. Amoebas slide over grains of sand hunting bacteria. Bacteria swim through micro-rivers in search of nutrients. Viruses attack bacteria, entering their bodies and using their protoplasm. Nematodes (a kind of worm) swim through the teeming forests of algae and microbes, eating almost anything that lives.

In the ideal soil community for sewage treatment, this microscopic matrix of living creatures recycles nutrients and water and transforms disease-causing organisms into harmless protoplasm. The tiny creatures bulldoze and devour organic debris that clogs the open spaces between soil particles. In the process, pore spaces open, cleaned water seeps downward, evaporates upward, and is utilized by higher plants.

This ideal kind of soil will have lots of air (oxygen) and be moist but not waterlogged. It is called an *aerobic* ("with air") community. The soil community that has little oxygen works more slowly and less effectively. It is called *anaerobic* ("without air"). Anaerobic soils are either very dry and compact or waterlogged. Aerobic vs. anaerobic is the most important biological distinction in this book.

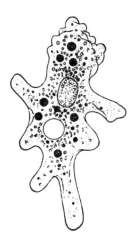

On the left, a nematode —just-visible creatures found in well-aerated soils. They eat protozoa like the amoebas (right) who, in turn, eat bacteria. Nematodes also prey directly on bacteria. In sewage treatment, their presence indicates aerobic conditions. Nematodes keep soil pores open by eating and moving.

The Aerobic Community

When soil is well ventilated with open pores and spaces, oxygen dissolves in the soil's water. The more dissolved oxygen, the faster and more completely these living creatures can "burn" (metabolize) organic matter. In soil sewage treatment, the organic matter is feces and other household wastes. The crucial living creatures are bacteria.

When oxygen is plentiful, species of bacteria (especially adapted to aerated and moist soils) feast, fatten, and multiply rapidly. They form thick, gelatinous colonies in the soil. As these aerobic-adapted bacteria consume and digest your feces and kitchen discards, heat is released. Because of the available oxygen, digestion is so thorough that organic "wastes" are completely "cooked" and chemically oxidized. When the aerobic bacteria excrete their "feces," they excrete stable and soluble chemical compounds. These compounds are like ashes in the fireplace—they need no more oxygen to digest them further. In other words, in aerobic soils, organic waste (not oxygen) limits bacterial growth. Competition is for your "wastes." Oxygen is not the limitation.

AEROBIC

O_2 + [organic matter] $\xrightarrow{\text{enzyme}}$ Bacteria protoplasm $C_5H_7O_2N$ + CO_2 carb dioxide / H_2O water / NO_3 nitrates / SO_4 sulphates / PO_4 phosphates + ENERGY (600 - 800 Kcal)

Between bacteria and higher plants are many other creatures that enjoy the benefits of well-aerated soils. For instance, protozoa are single-celled organisms that prey on bacteria. They can eat up to one million bacteria a day, but probably eat about 100,000 a day under typical conditions. The consumption of bacteria by protozoa is important because dormant bacteria (inactive in times of little food) and dead bacteria can pile up in soil pores. This pile-up prevents circulation of air through the pores. This may seem incredible until you think that a quart of wastewater may have 100,000 million bacterial bodies. In short, a good predator/prey balance keeps air circulating within the soil.

Instead of the microbial view, think how higher plants live in the aerobic community. Much of aerobic bacterial excrement is nutrient to higher plants. These nutrients are more soluble in water and more easily absorbed by plants than comparable chemicals excreted by anaerobic bacteria. Water and nutrients travel up, out of the ground into the stems and leaves. Here, part of the water is transpired from the breathing pores and a smaller part evaporates from the leaf surface. Up to 20% of the water in a soil (in a humid-temperate region) is absorbed *and returned* to the atmosphere by ordinary crops. This process, called *evapotranspiration,* is one of the primary ways wet soils are drained and water returned to the earth's water cycle.

Evapotranspiration is strong in aerobic communities. The heat released by bacterial digestion actually warms the soil. The increased warmth as well as moisture and aeration nurtures root growth. Increased roots bring even more air in the soil. A delicate feedback cycle is struck between soil warmth, soil bacteria, root growth, soil moisture, leaf shade, aeration and the sun.

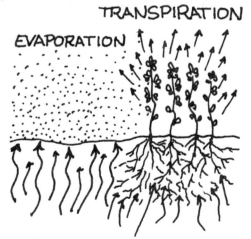

Stomata are small openings in the skin of leaves. This honeycomb lattice of air spaces within the leaf takes up about 1 per cent of the total surface. But more than 90 per cent of the water given off by the plant transpires through stomata.

The Anaerobic Community

Without enough oxygen, the aerobic bacteria and other soil microbes die or become dormant. In turn, their predators (like protozoa), who need lots of energy to hunt bacteria, die or become dormant. When oxygen is used faster than it can diffuse through the soil, *anaerobic* community of bacteria, yeasts, fungi, and actinomycetes takes over.

The anaerobic community is much slower than the aerobic in converting organic matter to soluble nutrients (it takes ten to twenty times as long). The anaerobic community also produces only one-tenth to one-fifteenth as much heat (energy) as is produced by aerobic metabolism. Anaerobic communities don't complete the combustion of organic matter like feces. Additional oxygen and other kinds of bacteria are needed to complete the process. Without oxygen, the whole community suffers: higher plants grow less and transpire less, and water movement is hampered.

Specifically, nitrogen, carbohydrates, sulfur, iron, and manganese are all changed to entirely different compounds from their forms under well-aerated conditions. Carbohydrates, for example, change to acids—not sugars. These acids, under extreme conditions, accumulate and can become toxic to higher plants. Nitrogen compounds change to "marsh gas" (methane and ammonia)—not to nitrates. Nitrates are the form of nitrogen most usable by plants. Consequently, roots do not get enough from the soil. Nutrient recycling and evapotranspiration are reduced in anaerobic soil conditions.

Finally, without dissolved oxygen iron combines with sulfur (instead of with oxygen). The resultant compound, ferrous sulfide, is black and insoluble. Ferrous sulfide combines with dead bacteria and algae to form a black gum or slime that clogs the pores of the soil. Soil drainage becomes drastically reduced. If this black gum, called the *organic mat,* spreads throughout the soil and if the soil does not aerate periodically, the drainfield may ultimately "fail."

$$Fe^{++} + S^{--} \longrightarrow FeS$$
$$\text{Iron Sulfide}$$

In summary, most soils have aerobic communities with pockets of anaerobic life. The anaerobic communities expand when oxygen supply is depleted. Keeping soils well aerated maintains the aerobic community and helps drainage. Providing air is one of the most important principles of drainfield design.

Water Movement in Soils

When water moves through soils, it brings nutrients to plants, brings oxygen to all the soil creatures, and purified wastewater by straining and filtering it through the soil mesh. We have seen that the "mesh" can be partially or even completely clogged *biologically* by the organic mat. But the mesh is also more open or closed depending on the *physical* arrangement of soil pore spaces, their numbers, and sizes. Soil pore spaces vary with the size and clumping patterns of the soil particles. Together, the organic mat and the open spaces control the movement of water through soil.

Biological Restraints on Water Movement:
Infiltration and the Organic Mat

Picture wastewater entering a hole. At first, it covers the bottom. It passes through the air/soil interface and then penetrates inside.

The ability of water to pass *through* the surface (*into* the soil) is called *infiltration.* Infiltration is very different from movement *inside* or *within* a soil. Passage of water *within* a soil is called *percolation.*

As water infiltrates into the soil, the pore spaces fill with water (replacing air). If the water keeps coming, the soil becomes saturated and no air is left (A). As soon as little or no air is available, the organic mat starts to accumulate. Since the bottom of a hole is saturated first, the organic mat starts there. The bottom surface layer of the hole becomes less and less pervious to the incoming water. As the bottom seals, the hole begins to fill and water infiltrates through the sidewalls only (B).

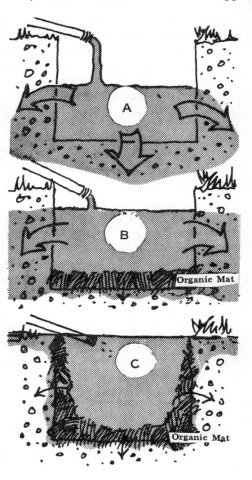

As the hole fills, micro-erosion occurs. Dirt particles slough off the walls, sink, and seal the bottom even more. If water keeps coming, the sidewalls begin to turn anaerobic. The organic mat spreads, and eventually the anaerobic mat covers the entire inside surface of the hole. Ultimately, if the hole is never given a chance to empty, the anaerobic mat could seal both the bottom and the sides of the hole. When the water can no longer infiltrate, it will surface above ground (C).

We have just detailed two fundamental principles of soil ecology and water movement. Soil clogging occurs *on the surface* and the organic mat makes *infiltration fail before percolation.* Many holes in the ground can be filled with standing water for days while the soil two feet away remains bone dry.

Sewage treatment in soils is basically a matter of maintaining infiltration in order to prevent clogging of the soil surface.

Physical Restraints on Water Movement:
The Void Space of Soils

Each kind of soil has beautiful and intricate 3-D patterns and arrangements of particles, clumps, and pores. The pores can be filled with air, water, or plants and animals. The 3-D network of pores (the void space) creates mini-aqueducts and underground rivulets for water transport and animal and plant movement. In some clay soils, the pores are so small that protozoa can't squeeze through the openings. In other soils, worms create the space by eating dirt and excreting it covered in a kind of glue. The gluey particles stick together to form the walls of earthworm tunnels, which in turn act as micro-tunnels for water to flow in. Other gravelly soils are so coarse that water rushes through the void space so quickly that plant roots can't absorb the nutrients. To help judge the ability of soils to transport water, a distinction is made between *soil texture* and *soil structure.*

Soil Texture

The texture of a soil comes from the proportions of its particles. We separate the particles into three groups, according to size: sand grains (the largest), then silt particles, and finally clay particles (the smallest). In general, the fine-textured soils are clays, clay loams, silt, and silt loams. They have small pores that slow down water movement. Sands, loamy sands, and sandy loams all have large pore spaces and transport water easily.*

---

* Humus is organic—consisting of decomposed leaves and animals. It can hold soil moisture because it is spongy. Humus is not considered in soil-texture analysis because it is found only in the top layers and varies so much from soil to soil. It is important for spray-irrigation sewage treatment, but it is not crucial to drainfields that irrigate underground.

Particles of mineral matter in soil.
This diagram shows them magnified
about fifty times.

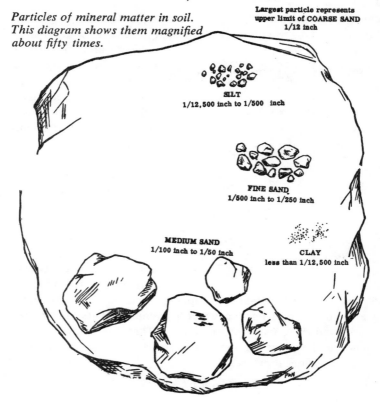

Largest particle represents
upper limit of COARSE SAND
1/12 inch

SILT
1/12,500 inch to 1/500 inch

FINE SAND
1/500 inch to 1/250 inch

MEDIUM SAND
1/100 inch to 1/50 inch

CLAY
less than 1/12,500 inch

Soil Structure

But soils are not just mixtures of sand, clay, and silt. Most soils come in chunks, crumbs, clods, irregular blocks, and jagged little slabs. These *aggregate* clumps occur because soil particles become cemented together by natural glues (silica, sesquioxides, and organic colloids). These cemented clumps of soil increase the ability of air and water (and earthworms and plant roots) to move through the soil. This clumping is called the *structure* of the soil.

There are two extremes of soil structure. Sand dunes, for example, have no structure, because all the sand grains are separate; there are no clumps. Old, overused agricultural land, on the other hand, can become a solid mass, with a very few randomly spaced cracks. This massive soil has no small clumps. Both the sand dune and the overused agricultural field are said to have no structure.

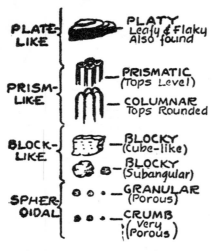

There are many kinds of clumping between the single-grain sand dunes and the overused agricultural land. Granular soil structures (having round clumps) and blocky structures (with cubelike clumps) allow water and air to move freely. Especially if the clumps do not overlap, the water moves quickly through mini-tunnels and mini-aqueducts formed by the clumps.

Platy soil structures (many flat plates) and blocky structures with a lot of overlap impede water movement. Water must find a path around these overlapping clumps. The water moves slowly and tends to back up. In short, soil structure is the main story in subsurface water movement.

The Soil Profile

All these variations in soil texture and soil structure came from millions of years of soil evolution. Rocks disintegrate, plants decompose, rivers flood and recede, oceans rise and fall leaving deposits of mud and sand, volcanoes and forest fires burn and chemically combine and restructure all the Earth's elements. These processes produce a layer cake of different soil types. The layering of one kind of soil on top of another is called the *soil profile*. Each layer differs in its texture, structure, and humus, its depth, slope, and color.

The best sewage treatment by soils occurs in thick soil layers that are well aerated and moist. The soil community is predominantly aerobic. The organic mat is confined to small anaerobic pockets. The soil structure and texture will allow easy passage of wastewater for filtering and purifying. Protozoa, nematodes, earthworms, and plant roots will thrive. Living creatures, soil, and water work harmoniously to recycle nutrients and return water and gases to the atmosphere.

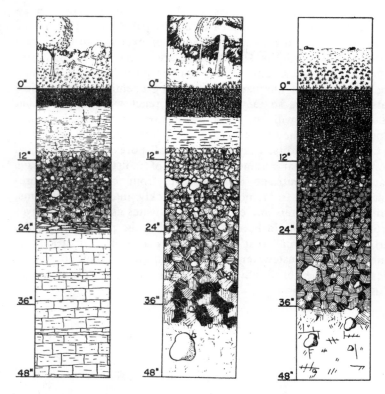

Soil profiles show the layers of different kinds of soils and bedrock. Knowing the soil profile is crucial to a good design of a drainfield. In general, soil has three layers: The top layer, with the most organic compost mixed with the mineral parts of the soil; the middle layer of weather-produced soil with little organic material; and a bottom layer of unweathered bedrock. The depth and kind of soil in the middle layer is most important to drainfield function.

ꜱꜱꜱꜱꜱꜱꜱꜱꜱꜱA THUMBNAIL SKETCH OF SOIL TYPESꜱꜱꜱꜱꜱꜱꜱꜱꜱꜱ

Clay particles are the smallest of all soil particles. They are so small that the total surface area of all the particles of clay in one pound of soil equals twenty-five acres of exposed surface area! Particles this small have electric charges that have significant influence on a soil's activity. The charged particles and great surface area perform numberless astounding chemical and physical feats: hold viruses so they can't move, react with nutrients and proteins so they will not wash away, precipitate chemicals like strontium 90 so they can't poison or pollute, and react with lime to change soil acidity. Clay, with its small pores and its chemical power, is the great retainer or holder of moisture. Plants have time to use the nutrients and the water from the soil. Clay prevents wastewater from moving too quickly into wells, streams, and aquifers. It filters and chemically renovates as the water moves through. But it should be remembered that the pore size and the particles are so small that *unstructured* clay soils can't transport water quickly. Wastewater may move so slowly that no oxygen circulates. Other clays have good, open structure. Water moves easily.

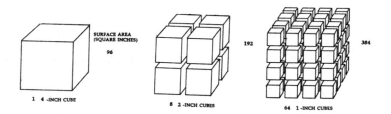

SURFACE AREA
(SQUARE INCHES)
96   192   384

1 4 -INCH CUBE   8 2 -INCH CUBES   64 1 -INCH CUBES

Although the total volume of the cube remains the same, the total surface area exposed to air increases as the cube is divided into smaller and smaller parts. The more surface area, the more cleaning power of the soils. One pound of topsoil has as much surface area as the whole state of Connecticut.

Silt particles are smaller than sand grains but larger than clay particles. Similarly, their pore spaces are smaller than sand pores but larger than clay pores. They can stop viruses and react with nutrients the way that clay does. Silt soils have lousy structure. They "melt" and can block drainage by filling in the cracks among rocks or gravel. Only in sandy soils do silts help by slowing drainage and giving the soil a better structure.†

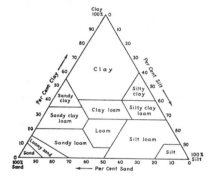

The Soil Conservation Service has developed this triangle to help define soil types. Each soil sample is put through a series of screens which separate out different sized particles. The percentage of particles of each size defines the soil. In many states, you can bring your soils to the SCS for free analysis.

Sands are the largest soil particles before gravel. They are chemically less active than clays. The surface area of a pound of sand is much less than that of clay (about three to five acres per pound of sand), but sand particles fit together in a way that results in large pore spaces between particles. Large pore spaces allow the soil to drain well, air to circulate in the soil, and protozoa to rush around and eat bacteria. A pure, coarse sand soil may be too porous for good sewage treatment. The effluent may move too rapidly to a stream or well and cause pollution. Fine sands filter bacteria, but coarse sands do not.

---

† A loam is a mixture of silt, sand, clay, and humus. A loam has about 40 per cent sand, 40 per cent silt, and 20 per cent clay. When there is lots of humus, we call it "topsoil." Sometimes, the word "silt" is misused to mean "loam."

A THUMBNAIL SKETCH OF

Viruses are the smallest pathogens—harmful organisms. In most soils, they have an electric charge. This charge comes from the protein coating surrounding the virus. The electric charge controls the virus's destiny because the virus is so extremely small. While moving through a soil in water, viruses are electrically attracted to clay particles. They "stick" to the surface of the particle. A pure clay soil can *ad*sorb ("sorb onto the surface") viruses within four *inches* of travel. Even in sandy soils, viruses are adsorbed in about one foot of travel!

A sewer project called the Santee Project wanted to use treated wastewater for swimming. Polio virus was poured into a natural aquifer of racks, sand, and only about 1 per cent clay. Two hundred feet later (at the first testing station) not one virus showed up. The ability of clay to adsorb viruses is tremendous. One hundred feet between drainfield and well (a typical safety margin) is obviously more than adequate.

Bacteria, as opposed to viruses, are rarely associated with sewage-caused disease. They are the simplest independent form of living creature. (Viruses are parasites on bacteria.) Bacteria eat pollutants, and their "waste" is the recyclable nutrients, water, gases, and energy that humans call good sewage treatment.

Pathogenic bacteria are weak outside the human body. They become easy prey for protozoa. Bacteria are much larger than viruses and are further removed from wastewater by straining and antibiotics produced by fungi and other creatures. The chance of bacteria escaping filtration through more than a few feet of any soil (except gravels or very coarse sand) is slight. Water-borne bacterial disease has never been traced to septic tank effluent in recent years.

MICROBE MOVEMENT IN SOIL

Health departments usually use the presence of one kind of bacteria (coliform) to spread health fears. ("High Coliform Counts Close River Bathing.") Know this:

—Coliform bacteria are not harmful. They are used to *indicate* pollution because they can be cheaply and easily identified. They are not "pollution" in themselves.
—Coliform bacteria *may* indicate human feces. But horses, cows, fish, ducks, even insects produce coliforms.
—Some harmless coliform bacteria grow wildly in lakes and rivers—totally exaggerating the illusion of pollution.
—One type (*fecal* coliform) is a better indicator of pollution, but even fecal coliform come from all warm-blooded creatures.
—What does the presence of coliform bacteria mean? *If* the coliform bacteria can be proven to be from humans, then *maybe* there are disease-causing organisms as well. (See Bibliography.)

Protozoa are the simplest animals. They eat bacteria. Some can adjust to pore spaces in the soil by changing shape and help keep the soil from becoming clogged with bacteria. To hunt well, they need oxygen and are much less effective in anaerobic soils. There are no important diseases caused by sewage protozoa in the United States.

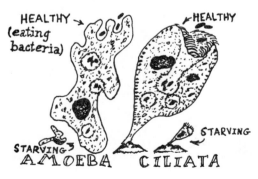

HEALTHY → (eating bacteria) ↗HEALTHY
STARVING STARVING
AMOEBA   CILIATA

Protozoa eliminate bacteria that might otherwise plug soil pores. The fat protozoa eat up to a million bacteria each day. The starving protozoa are staying alive on 10 to 20,-000.

The Soil Community and Human Health

The natural soil community eliminates disease-causing creatures (pathogens) in six ways: (1) it outcompetes pathogens for food; (2) the soil bacteria, fungi, and other microbes produce antibiotics that poison pathogens (penicillin is a soil mold); (3) the clay in soil adsorbs the viruses and keeps them from traveling to a place where they could reinfect humans; (4) the soil structure acts as a physical strainer to the larger bacteria; (5) the bacteria and many viruses in the bacteria are preyed upon by protozoa; and (6) the soil environment is so different from that inside the human body that most pathogens simply die from the radically different acidity, temperature, moisture, or shelter. Soils are so proficient that disease from septic-tank systems has never been proven in the United States. Health fears from home-site systems are totally exaggerated. This fear and other aspects of the myth of "dirty septic systems" have had terrible political consequence in the last twenty years of American life. They are discussed in the final chapter.

MILLIPEDE

4. THE IMPORTANCE OF SAVING WATER

Improving soil texture and soil aeration is expensive and, at times, the most difficult way to improve home-site sewage treatment. To replace one kind of soil with new soil requires excavating, trucking, and purchasing someone else's more porous earth. Adding air to the septic tank or drainfield requires pumps and more pipes and, sometimes, an outside, expensive source of energy for the pumps. On the other hand, water is easy to regulate—you can simply use less or more. Controlling water is the most economical way to reduce waste volumes. Water conservation helps everything from septic-tank/drainfield function to balancing the Earth's huge water cycle.

The Big Picture

Your body's plumbing (from mouth through intestines) is connected to your household plumbing through faucets, toilets, and drains. The sinks, showers, and toilets are, in turn, connected to the pipes and aqueducts of a utility company's water supply. And,

SEPTIC TANK PRACTICES

A GUIDE TO THE CONSERVATION and RE-USE OF HOUSEHOLD WASTEWATERS

finally, the water supply is connected to the reservoirs, lakes, rivers, and oceans of the earth. Modern Americans have become somewhat schizoid by blanking out the connection between their body's plumbing and the Earth's. Where does the water that comes from your faucet originate? Where does the water that you flush down the toilet go? In Los Angeles, I made a spot check by asking these questions and found that most people said, "Water? The city provides it." This kind of ignorance is more primitive and uncivilized than that of the Dark Ages of Europe. Like true schizophrenia, it is disconnected.

So, before considering septic-tank systems, stop and think more about the Big Picture. America continues the practice of doubling water supply by damming yet another wild river rather than halving consumption by reasonable water-conservation measures. Remember, the less water used and the more recycled, the less need for new dams and pipes as well as the bond issues and increased taxes that accompany construction. Water conservation can be accomplished by the small acts of many people in their homes. Household water conservation can replace the need for new waterworks and help preserve the integrity of small American communities as well as preserve natural lakes and free-flowing rivers. When you turn on the faucet, picture the network that binds your body to the biosphere.

The Water Cycle on the west coast of California. The sun causes evaporation. The clouds bring rain back to Earth. The rain becomes streams and rivers which flow back to the ocean. Ocean, streams, rivers and plants evaporate water back to the atmosphere. Plants and animals transpire water back.

On the left is the cover of an earlier edition. It shows how we connect water cycle to home life. Note that water carries nutrients from septic tank to feed crops.

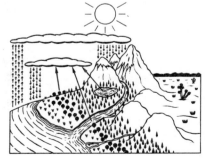

Septic-Tank/Drainfield Function and Water Use

It is obvious that the less water you use, the smaller (and less expensive) the septic tank and drainfield you will need. Even more important, a septic tank will treat sewage more completely when water use is low. First, the excrement from your toilet must remain in the septic tank to be treated. If water pours through continuously and rapidly, the tank overloads. Instead of settling to the bottom and being eaten by bacteria, large pieces of undigested feces may be pushed into the drainfield. The soils clog.

In addition, over the years soils become more and more packed as water rearranges particles and clumps. In a sense, water settles the soil particles and clumps in the same way that shaking a box of crackers compacts them toward the bottom. Obviously, the less water that passes through the system, the more porous the soil will remain and the longer the life of the drainfield.

Finally, we have seen how aeration and infiltration are closely connected. Cutting down on the amount of water passing through an already built drainfield will give the soil surfaces more time to aerate.

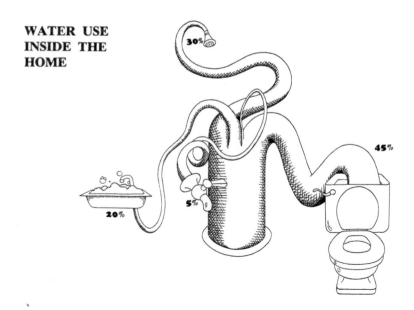

**WATER USE
INSIDE THE
HOME**

30%

45%

20%

5%

Present-Day Water Use in North America

A typical North American household uses between 50 and 60 gallons of water per person per day. A city household uses between 100 and 150 gallons per person per day (gpppd) if streets, restaurants, offices, cleaning, park watering, and fountains are included. The Public Health Service estimates 75 gpppd in designing septic tanks. But these are just averages. Some Native American tribes in the Southwest use only 5 gpppd, while some of the wealthiest communities—where there are garbage grinders and a multitude of other luxury appliances use up to 350 gpppd!

Water consumption decreases when the household water supply is metered—no flat rates. It increases with income (up to 50 per cent), garbage grinders (about 20 per cent), and washing machines (30 per cent). It decreases for low-income families (25 per cent or more).

The most extravagant use of water is flushing the toilet. About 45 per cent of a household's water (anywhere from 25 to 60 gpppd) is used to flush the toilet. The average North American family uses about 88,000 gallons of water each year; more than 35,000 gallons are just flushed. This extravagance is built into the modern flush toilet, which uses 5 to 7 gallons of good drinking water to flush about one-half pound of feces and/or around a pint of urine. After we mix such a huge volume of good water with such small quantities of wastes, the septic-tank/drainfield system tries to separate them—usually within moments after they were mixed!

NON-HOME WATER USE

We see only what we use. But we use much more than we see. It takes 1 to 2 tons of water to make a ton of bricks; 30 to 60 tons of water to refine each ton of petroleum; 250 tons of water for each ton of paper; 600 tons of water for each ton of nitrate fertilizer; 1,500 tons of water for each ton of wheat; 2,500 to 3,000 tons of water to produce one ton of synthetic rubber; 4,000 tons of water for each ton of rice, 10,000 tons of water for each ton of cotton. . . . Add this up and each of us uses 1,500 to 2,000 gallons every day to eat, drink, drive the car, and read the newspaper.

Obviously, the conventional flush toilet is costly and wasteful in areas where water is scarce. But even where water supply is not a problem, the flush toilet is costly and wasteful. Costly because you pay a utilities district to purify water to drinking standards (by chlorination and aeration) and then use the drinking water simply to carry away your wastes from your home. Costly because you pay to treat the carrier of feces and other wastes—not the wastes themselves.

Crapper's Valveless Water Waste Preventer.

(Patent No. 4,990.)

One moveable part only.

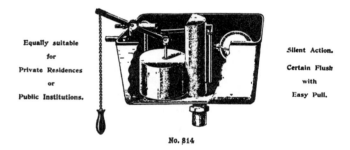

Equally suitable
for
Private Residences
or
Public Institutions.

Silent Action.

Certain Flush
with
Easy Pull.

No. 814

Quick and Powerful Discharge maintained throughout.

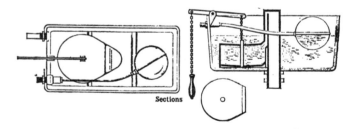

Sections

In the old days the water for a flushing toilet was provided from a cistern in which there was a valve at the outlet to the flush pipe. When you pulled the chain, it simply lifted up that valve and released the water. In other words you just pulled the plug out. Some people would

Household Water Conservation

To save wild rivers, stop pollution of lakes and oceans, save on taxes and bond issues, give a longer life to your septic-tank/drainfield system and let it work better, water can be conserved. Major household conservation measures are (1) changing habits and household practices, (2) installing devices and appliances that reduce water flow, and (3) water re-use.

---

tie the chain down so that the valve was perpetually open and the water flowed ceaselessly—either because they were too lazy to pull the chain every time or because they were ultrafastidious and wanted to ensure an immaculate flushing of the bowl.

This sort of thing horrified the Board of Trade, which used to be the ministry responsible for our water supply. They envisaged enough people doing it to cause all the reservoirs to dry up, and drought and pestilence could strike the land. But even worse was the second factor, although it would have seemed to have been of lesser importance. This was the fact that try as they might the makers of the valves could not ensure a snug fit. Each valve would start off watertight but it would not be long in use before it was failing to lodge properly after each flush.

The trickle, multiplied by thousands, was the Board of Trade's big worry. So the call went out for somebody to evolve a "Water Waste Preventer."

The trick was to make water flow uphill. If you consult the drawings opposite you will see how this was done. Or if you want to study it in the round, as it were, go to your nearest cistern and lift the cover and you will see precisely the same principle Thomas Crapper and his aides evolved at his Marlboro Works in Victorian times.

—From "Flushed with Pride: The Story of Thomas Crapper," by Wallace Reyburn, Prentice-Hall, Inc.

MATTERS OF HABIT

Avoid flushing the toilet for trivial reasons.
Use a wastepaper basket near the toilet for facial tissues and cigarette butts.

Don't flush after every urination unless there is a health or aesthetic compulsion.

Plug the bathroom sink when shaving or washing.
Turn off water while brushing teeth.

Don't use a garbage grinder. Compost garbage instead.
Garbage grinders use incredible amounts of water. The ground-up garbage heavily burdens the septic tank.

Use a dishwasher or clothes washer only when you have a full load.
Dishwashers use between 6 and 19 gallons a cycle. This does not include rinsing in the sink before loading. Clothes washers use from 40 to 60 gallons a cycle. Buy clothes washers with suds-savers and variable-water-level controls.

Be sure water isn't escaping from your toilet.
Check leaks by putting a few drops of food coloring in the tank of the toilet. Wait a few minutes or hours to see if color shows up in bowl. If it does, your ball valve needs replacing or the support assembly needs adjustment.

Plug the bathtub before starting water.
Fill only one-quarter full. Start with hot water only. Adjust later.

⬛AND HOUSEHOLD PRACTICE⬛

Limit your shower time. Turn off water when soaping up. Take a bath or shower with friends or relatives.

Unless you take a 4-to-5-minute shower, a bath will use less water.

Make immediate repairs on dripping faucets.

You can waste 12 gallons in 24 hours through drips.

Water the lawn or garden in the morning.

Evaporation is lowest then.

Use mulches around plants.

They save water as well as helping build soil.

Steam vegetables rather than boiling them.

Steaming uses less water, and leaves more flavor and food value.

Sweep sidewalks and driveways instead of washing them with a hose.

Remove ice trays and frozen foods from the freezer ahead of time.

You won't need water to loosen the cubes, and the frozen foods can thaw without running water over them. Keep a bottle of water in the refrigerator—it improves the taste by allowing the chlorine to escape, as well as saving water by not running the tap to get a cool drink.

Water-Saving Devices and Appliances

Aerators and Other Flow-Control Devices for Sinks and Showers

Sinks and showers can flow as fast as 8 to 12 gallons per minute. Aerators can reduce flows to 2.5 to 3.5 gallons per minute—a flow reduction of more than 50 per cent. There are new aerators entering the market that can reduce flows to 0.75 gallon per minute. Flow-control devices (usually plastic spindles or a metal fitting that fits into the pipe) save about the same amount as aerators. But *savings depend on your habits* (longer showers with flow control don't save water) and on the pressure in your water pipes. One study estimates that a flow-control device limiting flow to 3 gallons per minute in each family of four will save 7,300 gallons per year (about 4.5 per cent of the family's consumption). The shower flow control also saves on the cost of heating water—one of the largest expenses in every home.

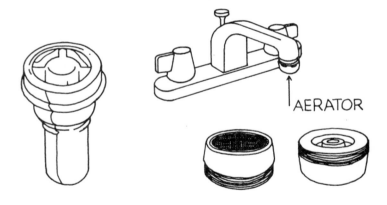

AERATOR

| A cheap and simple device to decrease water use in the shower by mixing water more thoroughly with air. See Noland in Manufacturers Distribution List in Appendix. | A faucet aerator can reduce water use by as much as one-third by mixing air with water as it leaves the tap. But, nothing is more effective than good water conservation habits. See Manufacturers List in Appendix. |

Instant Hot Water

Waiting for the water to heat up? That's a big waste. Try to minimize the distance between your water heater and the sink, shower, and bathtub—the shorter the run of pipe, the less cold water sitting in the pipes. *Insulate* the pipes—the cost is about a dollar per foot. This will save on energy costs of heating, as well as on water.

Clothes Washers

A typical front-loading clothes washer uses 20 gallons to wash, 40 gallons to spray-rinse, and another 20 gallons to deep-rinse. Depending on the style, a clothes washer uses from 40 to 60 gallons a wash. Two water-savers can be used with top-loading clothes washers: the suds-saver and the water-level control. The

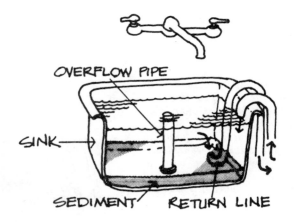

The suds-saver is a 20-gallon holding tank (like a service sink) located next to an automatic clothes washer. The wash water from the first load is stored in the suds-saver. The rinse cycles are unchanged. When the second load is placed in the washer, it is filled with the stored water. Three gallons are left in the sink with the sediment. Fresh water makes up the difference. The suds-saver (and the water-level control) are offered as built-in equipment in many major brands (see list of manufacturers).

suds-saver saves the wash water for the next wash. Up to 30 per cent of the water used can be saved, depending on the machine and on how many times you want to reuse the rinse water. The *water-level control* allows you to select the right amount of water for the size of the wash load.

As a rule of thumb, every pound of clothes requires between 1½ and 3 gallons of water. The water-level control usually has four selections: small load (10–12 pounds of clothes), medium (14 pounds), large (16 pounds), and extra large (18 pounds). Adjusting your machine to match your load will save lots of water. Don't be lazy, when buying an automatic washer get one with a suds-saver and a water-level control.

Front-loading clothes washers are reported to use one-third less water than top-loaders. It's best to compare pounds of clothes with water required for a complete cycle when buying a new machine. Front-loaders are becoming rare in the United States because of technical problems.

BRICK-IN-THE-TANK

CHERRY HILL, N.J. (Associated Press).—
Six months ago, Tilly Spetgong, a serious gal with a goofy idea, walked into city council carrying a brick. Councilman Steve Morgan ducked under his desk.
"He must have thought I was going to throw it," she said, "but all I wanted was to put one into every toilet tank in town." The unusual proposal to save water stunned the council, but it was approved.
The council anteed up $2,000 to buy 34,000 hardened bricks, the kind that won't break up in any kind of water and enough for every toilet in the town's 17,000 homes.
Last weekend, about 175 persons distributed 27,000 bricks, two to a house. They will finish this Saturday.
Mrs. Spetgong said: "If the average family of four flushes a total of 20 times a day we would save 34 million gallons of water every year in Cherry Hill."

Flush Reducers

If you already have a toilet that flushes 5 to 10 gallons with each use, then you need a flush reducer. The simplest is two standing hard-fired bricks. Instead of bricks, two plastic bottles weighted with pebbles can be used, or special dams or sleeves can be purchased to reduce the flush volume (see list of manufacturers). A whole series of newfangled contraptions that allow dual flushes (one for urine, one for feces) and restrictors of the chromium lever and combo sink-toilets have been patented (see *Residential Water Conservation* in Bibliography). In general, flush reducers could save between 10 and 15 per cent of total *indoor* water use each year. Even though this is half what manufacturers claim, it is still a huge portion of the family's water consumption. If you're metered, you will save money immediately.

If you are buying a new toilet, buy a low-flush toilet. They are slightly more expensive but price should go down as demand goes up.

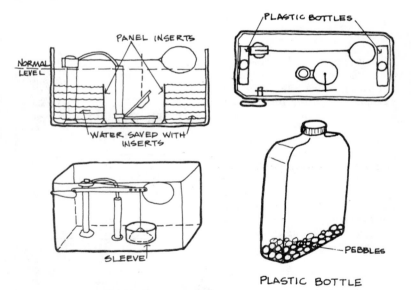

Besides bricks, plastic bottles, sleeves, or panel inserts can be used to reduce flushes. (Drawings from Residential Water Conservation, *see Bibliography.)*

Low-Flush Toilets

If you're buying, get a low-flush toilet. The most popular kind is the "shallow trap" water closet which costs about $10 more than regular toilets but uses only 3.5 gallons per flush (a 30–40% water savings). If you're metered, you save immediately. There are also dual-flush toilets with a low flush for urine and a big flush for feces. These devices are easily inserted into the toilet tank, except new flap-valve types.

A Japanese-made toilet with two flushes: one for urine and a larger flush for feces. Note that toilet tank is filled from a faucet. You can wash your hands and the wash water refills the toilet tank. Other toilets can be found in Residential Water Conservation *(see Bibliography).*

The pressurized toilet works by water flowing into empty tank and compressing the air inside the tank. When the force of air pressure and water pressure are equal, the water flow stops. When you depress the push button, the main valve is lifted inside tank. Tank water escapes into the bowl pushed by the compressed air and pulled by gravity. The drop in pressure causes the valve to close and the tank to refill in approximately one minute.

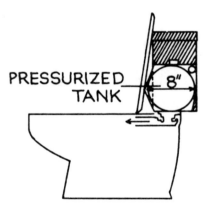

PRESSURIZED TANK 8"

Water Re-use

Water from your dishwasher, clothes washer, sinks, shower, and bath does not have as many disease-carrying organisms or pollutants as toilet wastewater does. The lightly polluted water is called "greywater," as opposed to the nitrogen-rich "black" water. The major failing of American sanitary codes, plumbing fixtures and designs, as well as water-conservation measures has been the lumping together of greywater and black water.

While water should be of very high quality for drinking, these standards can be somewhat reduced for bathing and cleaning, and greatly relaxed for irrigating and flushing. America's attitude toward water quality has become so confused that we use the greatest amount of top quality water for flushing—a water use that requires the least sanitary conditions. The most direct practical and responsible way to change America must finally come from government recognition that water quality should be related to its use and that codes and laws must distinguish between greywater and black water. These legal changes would encourage greywater recycling. Below are a few practical suggestions which may be illegal in your county.

Greywater is very variable depending on the home. 50 per cent of the phosphorus in a home presently comes from greywater, mainly detergents. Only 10 per cent of the nitrogen is in greywater. The other 90 per cent comes from urine and feces. According to one study (Monogram), only 2 per cent of the total coliform bacteria reside in greywater.

The Lavatory-Toilet Combination. This is the most direct way to recycle greywater for toilet flushing. As you wash your hands and face, the toilet tank fills. There are no health standards requiring high quality drinking water to merely flush . . . luckily.

Re-use of Greywater

Greywater recycling is already accepted for some uses in the United States. For instance, the "suds-saver" used with clothes washers recycles the first wash water for re-use with the second load. Commercial airlines use greywater (which they color blue) to flush the airplane's toilets. The Grand Canyon National Park has been recycling partially treated wastewater to flush toilets since the 1930s. The Campbell Soup Company is perhaps the largest recycler of "wastes" from their soup processing: The "waste" (a kind of industrial greywater) is used to grow more vegetables for Campbell's soup.

Greywater can be used to irrigate orchards, vineyards, lawns, fodder, fiber, and seed crops without health risks. These reuses allow little contact between humans and the greywater. Reuse for *edible* crops that are not processed (the way Campbell's soup vegetables are) is considered, at present, too risky. Similarly, irrigation of milk-cow or goat pastures and reuse for swimming, bathing, or washing the car (even if the greywater is filtered) are not yet acceptable, at least in California.

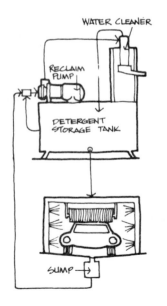

A recirculating carwash. Greywater recycling is commonly used in car washers even though no one has thought of the need for health regulations. Wash water is treated in a cyclone centrifuge which removes solid particles as sludge. Greywater is used to rinse cars.

Greywater Systems

Greywater recycling systems are new to America. They are in need of modern Rube Goldbergs and Benjamin Franklins. Greywater systems will vary with the *amount* of greywater produced and its intended re-use. In rural areas the amount of greywater used can be as low as 15 to 50 gallons per household per day. In suburban households, use may run from 50 to 200 gallons per household per day. City greywater use in one twenty-story apartment building can reach 20,000 gallons each day.

In most countries, greywater simply goes into the back yard. Open ditches direct the water to trees or crops that need watering, or the wash tub is carried to the plants and simply dumped. This direct re-use of greywater is called "the Mexican drain." This above-ground irrigation (without prior treatment) is banned in most parts of the United States—although it would save water and watering time.

Many of us living in drought-afflicted areas are experimenting with *above-ground* disposal of greywater. Methods include hay filters, sand filters, mini-lagoons, settling tanks like 55-gallon drums and swimming-pool filters with lawn sprayers. Health departments are still excessively concerned about pathogens in the kitchen sink. Although disease-spreading is possible, it is less likely than from coughing at the dinner table, kissing your best friend, failing to wash your hands before eating, changing the baby's diapers, or sharing food, wine, or cigarettes.

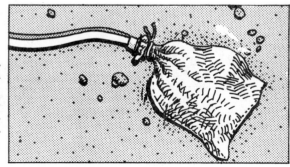

If you are running greywater straight into the garden, use a cloth bag at the end of the hose. Wash or change the cloth bag frequently.

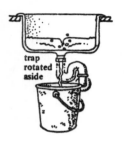

trap
rotated
aside

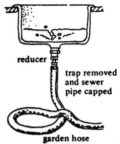

reducer

trap removed
and sewer
pipe capped

garden hose

*This is the "Mexican Drain"
—a hose direct from sink to
garden. The reducer will slow
the flow. If you want a permanent installation, install a
switch in the plumbing. You
can use the Mexican drain
in the summer and the drain-
pipes in the winter. A larger
size hose will let the greywater drain faster and reduce
clogging.*

*If you disconnect
drain (above), you
may want to cap
the outflow pipe
to prevent odors.
In the bathroom,
the collected grey-
water can be used
to flush the toilet.
This saves carry-
ing a heavy weight
(5 gallons weighs
42½ pounds).*

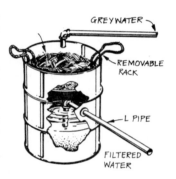

GREYWATER

REMOVABLE
RACK

L PIPE

FILTERED
WATER

*The RACK FILTER
has advantage of let-
ting greywater flow
quickly. It is not as
thorough a filter as
sand but is adequate for
most home greywater
re-use. Drum should
have screened cover. It
is best to let the grey-
water sit in the drum
and cool and separate
out the largest particles.
This can be done with
a plastic shut-off switch
(not illustrated).*

*The SAND FILTER
will slow down flow.
If clothes washer
greywater comes
pouring out, it may
overflow. So, fill the
drum only about ½
to ⅔ full with sand and gravel. Thirty
inches of sand above the gravel is plenty.
Sand filter shows grease trap if there is
kitchen greywater, fly-proofing screen.
The sand must be washed with clear
water every once in a while.*

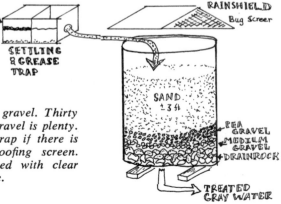

SETTLING
& GREASE
TRAP

RAINSHIELD
Bug Screen

SAND
± 3 ft

PEA
GRAVEL
MEDIUM
GRAVEL
DRAINROCK

TREATED
GRAY WATER

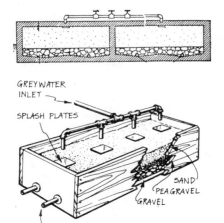

GREYWATER
INLET

SPLASH PLATES

SAND
PEA GRAVEL
GRAVEL

FILTERED WATER

THE BOX (of redwood or concrete) can handle larger flows than a single drum— although a series of drums side-by-side will serve the same purpose. Again, a cover is needed, especially to keep out rain which saturates sand and hinders filtration.

This sandbox design allows half of the filter to rest and to aerate—which reopens pores—while the other half works.

────────── GREYWATER CAUTIONS ──────────

• *Kitchen sinks contain grease in warm water. The grease must cool down to harden and then be removed by a grease trap—an ordinary filter will become clogged and rapidly produce odors. Some people only re-use bath, laundry, and lavatory greywater. They let both kitchen and toilet go to the septic tank.*
• *Do not use greywater on potted plants or seedlings unless you dilute it or alternate it with fresh water.*
• *Avoid using greywater on crops to be eaten raw, such as lettuce. On root crops, use both grey and fresh water to avoid any pollutant build-up.*
• *To avoid build-up of harmful ingredients, move greywater around garden. Don't leave it in one spot.*
• *Use on crops like tomatoes where edible part is not in contact with greywater. Fruit trees, artichokes, and ornamentals are also fine recipients of greywater.*

62   SEPTIC TANK PRACTICES

Below-ground disposal of greywater is legal and has few problems. The greywater usually goes through a mini-septic tank that acts as a grease trap for the kitchen sink and a lint trap for the laundry. Some friends with septic tanks use the septic tank for greywater during rainy seasons but have installed a valve so that they can irrigate with the greywater during dry times. When greywater systems are used with compost privies or other "dry toilets," the drainfield should be sized to handle all the greywater. If you waste fifty gallons of greywater each day, the drainfield should be sized for the whole fifty gallons. To soils, greywater and black water are not so different.

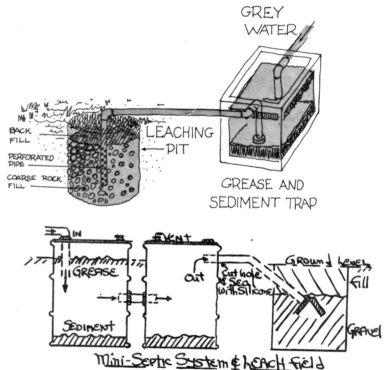

For small volumes, a miniature septic tank/drainfield can easily be built. This is a United Stand Design using coated 55-gallon drums and 1"×6" redwood plants to distribute water in the drainfield. For a state-of-the-art of greywater use, design, and health, see OAT volume listed in Bibliography or Greywater, *Box 15, Overlook, Bolinas, CA 94924; $2.*

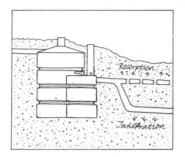

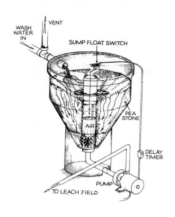

CIPAX (by Enviroscope, Inc., Corona del Mar, CA 92625) has a prebuilt design for about 80 gallons per day. This model has a dosing siphon and a two-compartment 200-gallon settling tank.

The Clivus greywater filter is expensive ($400). This luxury filter aerates as well as strains greywater. This process also equalizes the temperature which is useful for irrigation. The filter is well-tested and is already accepted in a few states such as Kentucky.

The most elaborate treatment for greywater re-use. Greywater first goes into settling tank where grease floats up and solids sink. Then it is pumped in doses to a sand filter. Finally, it is stored in holding tank and pumped when needed. This system can take 200 gallons per day.

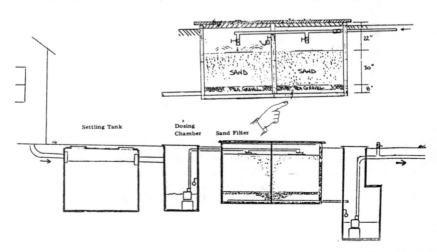

The Greywater Toilet

Re-use of greywater at home is not only possible—it saves up to 30 per cent of the total water used. In homes with greywater toilets *and* lawn sprinkling systems, even more water can be saved. The costs of building and operating the greywater toilet make it

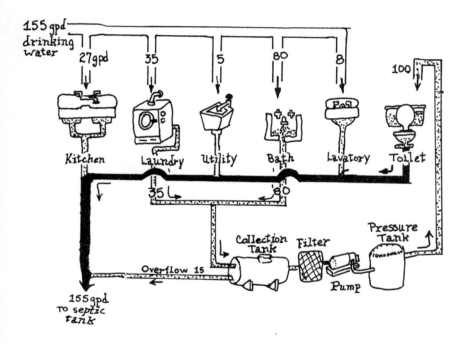

The flow from the kitchen sink (with its grease) and the toilet go to the septic tank. All other flush the toilet. For more information, see Demonstration of Waste Flow Reductions from Households *and Proceedings of the Conference on Water Conservation and Sewage Flow Reduction with Water-Saving Devices (listed in Bibliography.)*

most attractive where sewer and water rates are high, where multiple (condominium-type) housing is involved, and where septic-tank systems are in soils with poor drainage. This recycling helps septic-tank systems by reducing the loads, avoiding sudden party surges, and saving water. Here is the design for such a system (without the lawn sprinkler):

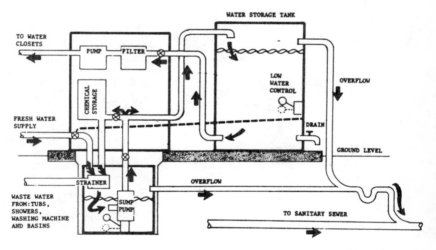

This is a commercial design to recycle greywater to toilets and for landscape watering. Cost, not available. Company: Aquasaver, 7920 Belair Road, Baltimore, MD 21236. Note that it has a fresh water "safety" in case there is not enough greywater generated to flush the toilet.

MANUFACTURERS AND DISTRIBUTORS OF WATER-CONSERVATION PRODUCTS: SEE APPENDIX 3.

City water use can be cut significantly if city councils and water boards are willing to change building codes and use their imaginations. In this ideal apartment house, fresh water is used for bathing, drinking, and clothing washing. The greywater produced is recycled from a storage tank in the cellar. It can be used to run vacuum toilets and water-cooled air conditioners. It can be re-used for street washing, roof gardens, skyscraper fire protection, parks and recreation lakes, and factory cooling. To achieve city water conservation, a new look at urban plumbing and piping is needed. It is too expensive to replumb the Empire State Building at this late date.

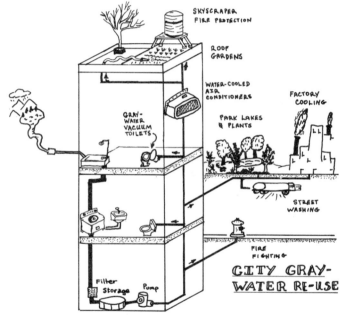

AVERAGE 20 FLOOR BUILDING

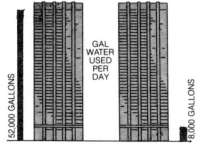

In the city water use can be curtailed by using vacuum toilets in apartments and business buildings. Vacuum toilets suck wastes through the pipe—requiring about three pints per flush. For information: Colt Industries, Water and Waste Management, 701 Lawton Ave., Beloit, WI 53511.

5. SEPTIC TANK DESIGN

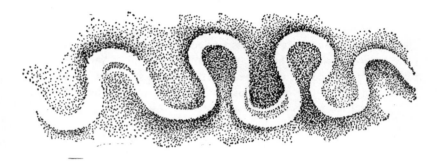

The best design to trap solids inside the septic tank is based on the meander of old rivers. These rivers move slowly and collect the most silt.

As people learned more and more about wastewater and soils, the design of the septic tank changed. A little before the turn of the century two Frenchmen, Mouras and Moigno, discovered that a big box (the "tank") placed between the house and the cesspool trapped the excrement, reduced the amount of solids, and produced a clarified liquid ("effluent") that entered the soil more quickly. In the 1960s, when the principles of soil clogging were studied, further modifications of the box were suggested. These newly designed boxes kept even more solids from entering the soil, gave more time for bacteria to eat the feces, and made it easier to remove the undigestible sludge that accumulates inside the box. At present the main ecological reason for having a septic tank is to protect the soil from clogging quickly. Maybe in the 1980s these boxes will be considered a home source for good fertilizer for the garden.

What Happens in a Septic Tank

The septic tank is, in some ways, a miniature ecosystem. Bacteria will flourish or struggle along depending on the water temperature, the amount of oxygen, the rhythm of water use, the chemicals used by the household, the acidity they create, and the family diet. For instance, bacteria like temperatures between 70° and 80° F (21° to 27° C). They slow down if the water temperature is much lower.

When household sewage enters the septic tank, the heavy particles settle downwards. Other suspended particles may coagulate to form heavy particles. These settle downwards at a slower rate. Between 70 to 80 per cent of the larger particles have settled to the bottom of a four-foot-deep tank within three hours. The accumulation of settled solids is called *sludge*.

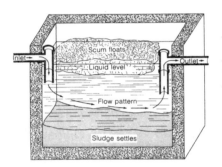

A one-compartment septic tank. This old-fashioned design is not as effective as multicompartment tanks. Some solids enter the drainfield and clog soils.

Meanwhile, substances lighter than water (like grease and fat) float upwards. Bubbles of gas, given off by well-fed bacteria, carry some of the settled solids back up to the surface. The combination of floatables is called *scum*.

Part of the scum and sludge is not edible by bacteria. This undigested part must be stored in the septic tank for removal by pumps or pails. So the septic tank not only gives bacteria enough time to reduce pollutants but also has enough room to store the indigestibles.

Give Microbes Food They Can Eat, Don't Poison Them

The micro-organisms in a septic tank can eat quite a varied diet but some substances are just impossible. Cigar and cigarette butts, filters, sanitary napkins, facial tissues, hair, paper towels and napkins, Band-Aids, old toothbrushes, and broken children's toys all belong in the garbage pail, not the toilet. They can't be eaten *and* they take up precious space in the septic tank.

The first septic tanks appeared in the United States about 1883, when a two-chamber, round, vertical tank equipped with a dosing siphon for discharge was designed by Edward S. Philbrick of Boston, Mass.

Grease and fat will not digest in a septic tank. Most will float to the top as scum, but grease will combine with laundry detergents and wash through the septic tank into the drainfield. There it clogs the soils. Put grease in a can (glass breaks).

Be wary of any product said to clean septic tanks. Many of these products temporarily precipitate solids—giving the illusion of success—but they cause a solid bulk that is more difficult for bacteria to eat, and change the water's acidity so much that many bacteria perish. Products with sodium or potassium hydroxide are the worst. They not only ruin long-term digestion but also significantly speed soil clogging.

There are more than 1,200 products asserting that their "enzymes" will help septic tank digestion. Bacteria produce their own enzymes and only eat as much as their *own* enzymes can digest. So far, no enzyme product has been proven beneficial.

Obviously, large amounts of any single chemical will hurt the septic tank ecosystem. Chemicals from darkrooms or large amounts of paints should be directed into a separate dry well.

Keep the Solids in the Tank

The solid "wastes" from your household must be given time to settle down or float up—so that they will not flow out into the drainfield. If the solids flow through the septic tank into the drainfield, the soil will clog quickly. There are many theories about how long household solids must sit in a septic tank for the predominantly anaerobic bacteria to eat as much as they can. Obviously, the longer, the better. At present it is believed that three to five days' storage will give bacteria sufficient time to digest a sizable portion of the solids.

There are three ways to keep solids in the septic tank longer: (1) reduce water consumption so less wastewater flows through the tank; (2) make the septic tank larger; (3) arrange the shape of the tank so that more scum and sludge float and sink out of the main flow. While codes have required increasingly larger (and more expensive) tanks, they have ignored water conservation, tank shape, and flow arrangement. Here are a few new design features and ways of estimating needed tank size.

Shape of the Flow

The main trick of septic tank design is to make it like an old, deep, winding river. Make it deep so the solids will settle out of the main flow. Make it wide, so the flow isn't tight and restricted and fast like a new-born stream, but not so wide that there will be corners with no flow at all (like an alluvial plain). Make the volume large, so the incoming wastewater flood can disperse its energy into the huge "lake" of the tank. (When the flow loses energy, the sediment sinks and air bubbles let the scum float up.)

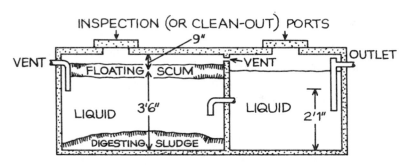

A two-compartment septic tank removes more solids than a single-compartment tank, but less than a three-compartment tank.

A series of three-cylinder tanks made from agricultural pipe removed more solids than any other combo. The study was done under the Public Health Service during the 1950s.

Give the tank compartments (whose small openings act like a river's water gaps). Each compartment lets only a small stream flow through it. All the rest of the suspended scum and sludge hit a wall and float up or settle down. Finally, make the wastewater river reverse its direction of flow again and again. This "meander" makes the river slow down and, like the old Mississippi, drop its sediment at each bend.

In the 1960s Dr. J. T. Winneberger perfected the three-compartment, meander tank. Deep, large, and divided into three compartments in a meander-flow design, this tank retains more solids and produces a clearer effluent than all other designs.

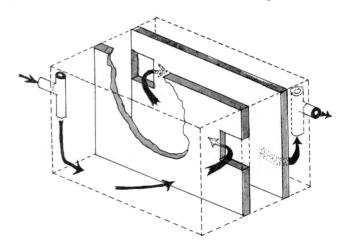

The three-compartment meander tank removes more solids by slowing flow and subdividing tanks into many chambers.

Septic Tank Size

Nondigestible solids accumulate in fickle ways. One kind of family diet will lead to lots of sludge, another to lots of scum. Sludge and scum become compacted over the years and some of the "indigestibles" finally break down; so, over the years, less and less sludge and scum appear to accumulate. Because septic tank storage is so erratic, it is impossible to recommend a specific size for, say, five years' storage. With no scientific way to size a septic tank, we make semiknowledgeable guesstimates.

We do know that 70 to 80 per cent of the solids in a septic tank are digested within three to five days. Therefore, the septic tank should be able to hold at least three days' volume of wastewater. For example, a family of four wasting 75 gallons of water per person per day (300 gallons for the whole family) might buy a tank with a 900-gallon capacity. But the longer the solids sit, the better. The family of four could reduce their water consumption. If they used only 50 gallons per person per day (200 gallons for the family), a 600-gallon tank might be plenty. But the longer the solids

Inlet and outlet baffles prevent sludge and scum from backing up into the household plumbing and flowing into the drainfield. They come in all shapes and materials. At left is a tee. Center is an ell. At right is a type found in old metal tanks and some new concrete tanks. The current best buys are plastic tees, since they allow easy cleaning from the top. If agricultural pipe is available, it works just fine, too.

sit in the tank, the more will be digested and the clearer the waste-water entering the drainfield. The same family might decide to use less water and the bigger tank. The solids would then sit in the tank four and a half days instead of three.

There is nothing sacred about these numbers. Size is best estimated by water use and balanced by cost. If it doesn't cost too much, get a larger tank, but don't get a larger tank as an excuse to waste more water. Remember, it may be cheaper to have a grey-water dry well, a smaller septic tank, and a smaller drainfield.

Under present-day construction practices, most new homes have a septic tank with a capacity between 900 and 1,200 gallons.

Summary

The septic tank has been improved in shape with compartments, baffles, and manholes.* The size and design allow solids to settle or to float out of the wastewater, to be eaten by microbes, and to store sludge and scum for a reasonable length of time. Keeping water use low and indigestible wastes (like cigarette filters) from being flushed down the toilet cannot be designed into the septic tank. They require the understanding and thought of the household. Please, care for your microbes and they will respond in kind.

---

* This chapter does not cover certain features of septic tank design because they are obvious, but they do require attention when installing or renovating an old tank. Items such as manhole covers, the materials and construction of the septic tank, and the kinds of baffles to prevent solids from moving into the drainfield are described in chapter 8.

6. DRAINFIELD DESIGN

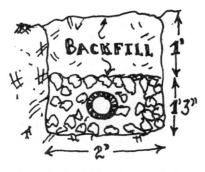

A typical old-fashioned drain-field trench. It functions poorly because it relies on bottom area which clogs quickly. This trench is still standard in several states —e.g., Kentucky. Other problems include gravel size and pipe placement.

This is an ideal trench design that allows wastewater to infiltrate the surrounding soil easily and quickly. It uses sidewall areas (rather than bottom) for maximum infiltration. See chapter 8 for construction details.

The drainfield soil filters, strains, and chemically renovates the polluted effluent coming from the septic tank. Soil is nature's astounding purifier. The best drainfield will let effluent infiltrate into the soil and percolate away year after year. Since infiltration is the first aspect of water transport to fail, drainfield design is based on maintaining the infiltrative surface—keeping it from clogging.

Follow these three guidelines to help keep the soil infiltrating:

1. Use as little water as possible.
2. Dig the drainfield in the best soil layers for water movement.
3. Encourage the aerobic soil community in the drainfield.

The guidelines interact: If you use little water, then the surfaces of the hole get more air. If you use the more pervious soil layers, then the cleaned water will move away faster and the walls of the drainfield hole will again receive more aeration. Aeration, infiltration, and percolation mutually affect each other.

The Drainfield Hole

Before 1926 Western Man improvised all kinds of holes in the ground for the effluent coming from the septic tank. In 1926 Henry Ryon, a sanitary engineer working for the State of New York, became fed up. Too many complaints came into his office each day. Henry Ryon became the first to consider how soils provide sewage treatment. He was followed by a group of scientists at the University of California who investigated soil clogging. They added to Ryon's knowledge and recommended that the drainfield hole have a specific shape and depth.

Sidewalls Only

The scientists found that the bottom of any drainfield hole is quickly saturated with effluent. The organic mat grows quickly in the saturated bottom and clogs the soil. In short, the bottom is not an infiltrative surface for long. Only the sidewalls are effective, because they receive some air as effluent levels go up and down. In addition, the sidewalls erode, stripping off the organic mat. The shape of any drainfield hole should emphasize the sides—not the bottom area.*

This recommendation is contrary to the U. S. Public Health Service's *Manual of Septic Tank Practice,* which says soil absorption area in the hole "is figured as the trench *bottom* area, and in-

---

* An exception to this rule occurs when evapotranspiration rather than draining becomes the most important method of removing water. In this case a shallow bed that allows root penetration is preferable. Such situations occur in the Southwest (see the next chapter for details).

cludes statistical allowance for the vertical sidewall areas." This is ecological nonsense and should not be followed. Only the sidewalls are effective infiltration surfaces.

The Best Soil Layers

Try to use the layers of soil that will transport water away quickly. Obviously, avoid putting the drainfield hole in heavy clay, which impedes water transport and aeration. The soil profile exposes the best layers for water transport. If, for instance, there is a layer of clay and then, below it, a layer of sand, it is best to dig the hole deep into the sand. On the other hand, if the soil is all sandy loam, then the depth and sidewall shape of the hole are not critical. There is no one best sidewall shape or depth. The shape and depth depend on your local soils. Sometimes narrow trenches are best. Sometimes deep seepage pits work better. Look at the soil layers before you design.

A poorly placed trench with sidewalls in clay layer.

A well-placed trench using sandy layer for part of its infiltration area.

Avoid Groundwater

The shape and depth of the drainfield hole(s) are influenced by the big geological and water picture. The most common concern is the combined effect of an impermeable layer of rock or soil plus heavy rains. If the impermeable layer is near the surface, water "perches" or sits on the solid rock or hardpan. The groundwater level rises and begins to flood the drainfield. *Sometimes* this causes anaerobic conditions and the growth of the organic mat.

There are lots of superstitions about impermeable layers and groundwater. Some drainfields work totally flooded in water. Others fail when the groundwater is within two feet of the drainfield's bottom. It is best to avoid flooding your drainfield. But there is no way of specifying one depth that is right without knowing the overall picture of underground water movement, the degree of flooding, and the period of continual flooding.

Every authority in North America has advice on how far the bottom of the drainfield hole should be from groundwater or bedrock; some say four feet, some eight, and others say the bottom itself. These "magic" numbers, which don't take into account local geology or water tables, have become punitive rules discouraging septic-tank systems in many parts of North America. To repeat, excavate the drainfield to avoid flooding by perched water. But also remember that some drainfields work even under temporarily flooded conditions. Ask your neighbors about their rainy-season experiences.

A trench may be flooded by water that "perches" on a rock layer. Permanently flooded trenches may cause anaerobic conditions in poor soils, especially if the condition lasts for many months. A fluctuating water table does not permanently flood the drainfield; there is temporary aeration. Temporary flooding is tolerable if the septic tank is high enough to force effluent into the drainfield, preventing sluggish drains.

Fractured Rock WITH Soil

Some geologists and health officials fear that wastewater travels too fast and is not cleaned in *fractured* bedrock and *gravelly* soils. Polluted water could arrive at a well or stream. These fears are most often unjustified. Neither fractured bedrock nor coarse soils are *ipso facto* bad for sewage treatment. Some coarse soils with just 1 per cent clay will provide adequate filtration (see Thumbnail sketch of Soil Types, in chapter 3). In fractured bedrock (with some cracks filled with soil) water transport and sewage treatment are superb. Only a few fractured bedrocks with absolutely no soil in the cracks *might* be dangerous. To be safe, if the soil seems too pervious, try to dig the drainfield where the water must travel through some finer soil, or import some finer soil (chapter 7).

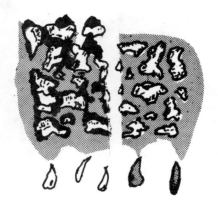

Left, wastewater passing through rock with some soil. Wastewater can easily be treated if passage is more than 100 feet.
Right, wastewater passing through rock without any soil. Wastewater is not treated.

Summary

Sidewalls, the soil profile, hardpans and solid bedrock, coarse gravels and fractured bedrock are the environmental guides to drainfield shape and depth. The best shapes are narrow trenches and small-diameter pits.

The Size of the Drainfield

Ecologically, the size of a drainfield is determined by the long-term ability of your soil to let wastewater infiltrate and the amount of wastewater that must infiltrate the soil each day. Obviously, the better the long-term infiltration and the less the waste-water, the smaller the drainfield need be. Economically, the size of the drainfield depends on the cost of hiring someone and his machine to excavate the hole. Politically, you may feel you must follow the local codes—no matter how inappropriate their sizing procedures.

At present there is a lot of confusion about sizing a drainfield. No scientist, engineer, or layman can predict long-term soil infiltration. There is no "test" that can give anything but crude "guesstimates." These guesstimates are based on the percolation test, soil texture analysis, and the advice of your neighbors. Even more than septic tank size, appropriate drainfield size has more in common with horse betting than science.

The Percolation Test

The "percolation test" is actually misnamed, because it measures both infiltration and percolation. It consists of digging a very small hole, pouring in water, and measuring how fast the water disappears.† The test is very fickle. The way the hole is dug, how long the surrounding soil is soaked, the shape of the hole, and the way the water drop is measured can all change the rate of water drop. One professional will say the water disappeared quickly; another, equally competent, says too slowly.

In addition, the relation between an *overnight* test and long-term soil receptivity is hard to determine. If the water in the hole doesn't disappear, you can be sure that there will be problems with aeration and the organic mat. But if the water disappears in an hour or so, you still don't know if it will *always* disappear in an hour or so. Maybe after a year's use the water will take two hours, or half an hour. Nobody can say.

† Precise details of the percolation test are given in chapter 8.

A very general chart of how fast a gallon of water might disappear into a square foot of a particular soil surface has been worked out by Henry Ryon and updated (by eliminating bottom area) by Dr. J. T. Winneberger. This chart simply comes from their experience and is a very rough estimate. You can see, for instance, that if it takes three minutes for the water to drop one inch, you might use 1.6 square feet of *sidewall* surface for each gallon of household water.

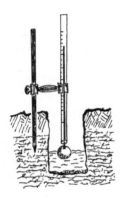

The percolation test to measure the rate of water drop in a hole. Clamp holds a plastic tube that has an inch scale on it. A plastic rod attached to a styrofoam ball moves up and down inside the tube. Details in chapter 8.

Ryon's Corrected Criteria, Maximum Wastewater Loading Rates for Drainfields

Percolation Rate Min./in.	Maximum Loading Rate gal./sq. ft.*/day
1 or less	2.5
2	2.0
3	1.6
4	1.4
5	1.3
10	1.0
20	0.72
30	0.48
40	0.42
60	0.36

* Active infiltration surfaces are sidewalls of disposal fields.

The Soil-Texture Test

The soil-texture test does not measure infiltration or water movement of any kind. It simply analyzes the soil for the amounts of clay, silt, and sand. The sandier the soil, the better the soil is for water movement—so the story is told. Of course, the test by itself is very crude. Infiltration is a combination of soil texture *plus* soil structure, aerobics, the kind of clay, and all the other things discussed in chapter 4.

Jack Abney (see chart) extrapolated his experience of soil texture and how many gallons of water would infiltrate through a square foot of soil. Pretty sloppy, but if you can affort a soil-texture analysis *and* a perc test, you will begin to get a mystic feeling about the amount of the minimal of sidewall area you need.‡

See page 93: Soil Colors Give
Helpful Hints.
See page 106: Soil Conservation
Service Gives Helpful Advice.

Abney's Loading Rates

Soil Texture	Sidewall for Each Gallon
Clay/silt (less than 0.1 mm)	4 sq. ft.
Sandy soils (greater than 0.1 mm; less than 1.0 mm)	2.5 sq. ft.
Coarse sands,* small gravel (greater than 1.0 mm)	1.5 to 2 sq. ft.

* When soils are this coarse, check drainage to make sure no nutrients or pathogens can pass through them.

---

‡ Details of the soil texture test are given in chapter 8.

Neighbors and Other Humans

If your neighbors' drainfield works, that is the best indication that yours will too. Check to see if the soils are similar. Ask about problems, how much water the neighbors use; the size of their drainfield. Nothing beats their experience (and an honest local contractor's). They might just say, "Oh *my* drainfield's half the size and it's worked for forty years."

Obviously, if you can afford the percolation test (or five) *and* the soil-texture analysis—*and* talk to your neighbors, local contractor, and local Soil Conservation Service man—you're doing the best possible. All these people will have different opinions. Some, especially the professional "hard" advice, is probably nonsense. If you want a drainfield with a long life, *excavate as much sidewall area as you can easily afford,* using these tests as guides. But don't be pressed into superexpensive holes just because the codes say so. If you feel you will use significantly less water than the codes estimate, or you feel they are not using the percolation test or soil texture tests appropriately, go before the appeal board. Use this book. Every code in the United States has to have an appeal process. Most times, if you use a little intelligence, the appeal board will give you the go-ahead.

Two Popular But Terrible Methods to Size a Drainfield

Amazingly, the **number of bedrooms** *in the house is rapidly becoming the most popular "standard" used by local health departments to determine drainfield design. These estimates are obviously ridiculous. Bedrooms do not waste water—people do. At one end of the American spectrum are communal life-styles and Native American homes where everybody essentially shares one large room. At the other end is the middle-class ideal where each individual has his/her own bedroom (kids too!). Bedrooms, according to this crazy notion, all waste water into identical soils.*

The **average water use,** *based on somebody's idea of what a typical American household must waste each day, is also used to size a drainfield. Most often this standard assumes you are a slob and waste water extravagantly. The drainfield usually ends up bigger than needed— which means it may last longer, but which also is punitive: It makes conservation-minded homes pay more than necessary for their water-saving ethics.*

Drainfield Layout on Flat Ground

On flat ground, the drainfield can have any layout that allows wastewater from the septic tank to flow into the total excavation. The interconnected trenches (or pits) let the wastewater spread throughout the whole drainfield until it reaches the same level as inside the tank. There are many variations for flat-ground layouts (forklike, serpentine, or grid). The layout design does not matter as long as the ground is level. On slopes, these flat-ground layouts don't work. They distribute wastewater unevenly, so that one section is flooded while another remains dry. The flooded section will become anaerobic and clog before the others. The highest trench on a slope may, in fact, be empty while the bottom is overflowing. The only way to prevent this unequal distribution is to divide the serpentlike trench into separate units and make certain (by pipe arrangement) that the first is totally full before the second, the second before the third, etc. This slope layout is called *serial distribution* and will be described in the next chapter.

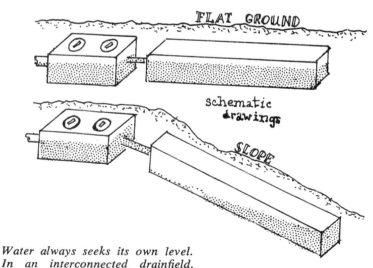

Water always seeks its own level.
In an interconnected drainfield,
wastewaters spread out until the whole drainfield is filled to tank level. Illustration shows how all the effluent rushes to the bottom, causing clogging. The next chapter shows how to design a trench system that solves this problem.

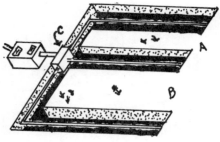

The *FORKLIKE LAYOUT* has this advantage: If one of the "tines" should clog or a truck run it over and crush it, the other "tines" would work. It has the disadvantage of wasting useful infiltration area ("A" and "B"). The layout sketched at left uses a distribution box ("C"). (See explanation on page 86.) If clogging occurs in the distribution box, one trench may never get used.

The *RECTANGULAR* or *GRID LAYOUT* has the advantages of using the soil area most efficiently and still using all the trenches if a break occurs. The grid drainfield cannot be used in an interlaced dual drainfield.

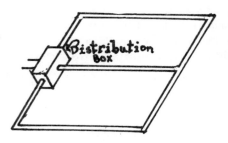

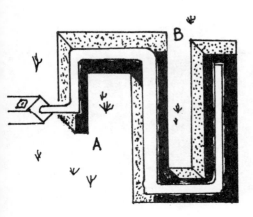

The *SERPENTLIKE LAYOUT* has the advantage of being able to wander to where you want it. If you are subirrigating with the wastewaters, you can lay out the pipes near the crops—rather than put the crops where the trenches are. Disadvantages: wasted infiltration area ("A" and "B") and a break or clog toward the beginning of the trench will prevent the rest of the trench from being used.

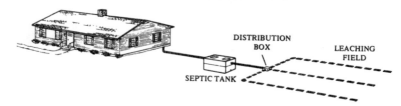

The distribution box is a chamber which shunts effluent into the drainfield pipes. If easily accessible (it should be), the distribution box is also an inspection manhole to see if any large solids are entering the drainfield. Distribution boxes work well on flat land but, on slopes, tend to let effluent go out only one of the many outlets. Distribution boxes should be checked yearly to make sure one of the outlets is not clogged. The clogged trench will be dry, while the others are overloaded.

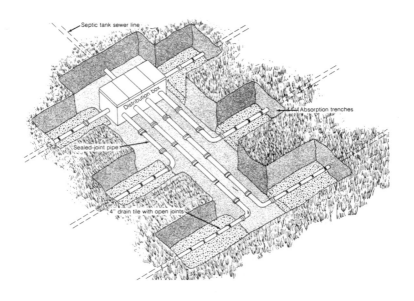

Soil Aeration, Dual Drainfields, and Long Life

To assure that air gets into the soils, engineers have recently invented a drainfield design that allows much greater aeration. This is called the *dual drainfield* or *alternate bed system*.

Two drainfields are built instead of one. Each is a *complete* drainfield. The two drainfields are connected to the septic tank by a flow-control box or diversion valve. Each drainfield receives effluent for one year. Then the flow of effluent is switched to the other field and the first field gets a rest and a chance to aerate. The decomposed humus developed after several years of alternate loading and resting may even improve clay soils and help sandy soils hold nutrients.

Details of the flow-control box and how to arrange a dual drainfield system are given in chapter 8. These boxes are commercially available, but they can be made easily.

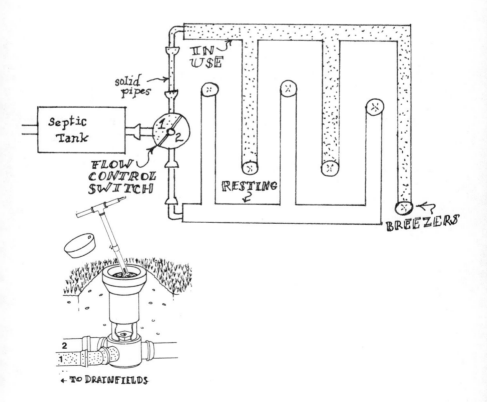

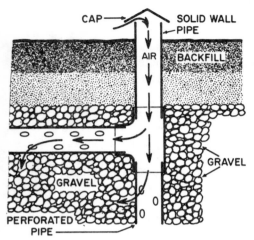

Breezers

Daily aeration can be improved by the use of breezers or breathers. A breezer is a pipe perforated at the bottom and open to the air on top. A breeze enters the pipe and pushes air into the trench. Breezers function best when placed at the end of each trench or in each connected seepage pit.

Summary

Drainfield design and function improve when you use as little water as possible, encourage aeration of the soil, and dig the drainfield in the best soil layers for water transport. These three principles have been understood only recently. Most codes and manuals are unaware of soil, air, and water interactions on drainfield design.

Only the major aspects of drainfield design have been outlined. In addition, we have assumed basically flat ground with good drainage for drainfield construction. In the next two chapters we will consider problem soils, slopes, and details of construction.

7. ECOLOGICAL PROBLEMS

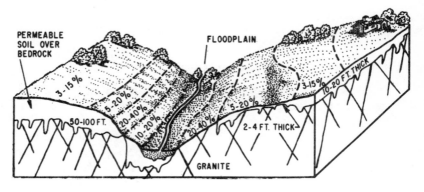

Slopes, soils, and bedrock must all be considered when planning a drainfield. Slopes of more than 20 per cent are hard to dig on. Both sides of the creek are good septic-tank/drainfield habitats in the 3–15 per cent slope area. The left-hand side is good between 5 and 20 per cent, but the right-hand side is questionable with normal drainfields. Most soils that are 2 feet thick or less need special man-created drainfields. Flood plains are difficult for all kinds of sewers—on-site or centralized. They are best avoided.

In some situations septic-tank systems may be difficult to install and/or operate. The topography may be unsuitable because the slopes are steep. The house may be located in the flood plain of a creek. The soils may be impermeable or have one layer (called a "pan" or "fragipan") that is impermeable. The soils may be too shallow (for example, three feet to the bedrock). Or the water table may be too high for too much of the year—preventing air from entering the drainfield.

When installation of a septic-tank system seems questionable, think of the alternatives: maybe a compost privy or a system that recycles most of the water. When septic-tank systems are difficult to use, there is usually so much water that the wastewater can't be treated. The easiest solution is a waterless toilet and a grey-water recycling system.

If there are soil and water-table problems, the septic-tank system will cost more for construction (and maybe operation) than a simple drainfield in well-suited soils. Except for the dosing tank, the solutions suggested below do not use additional electrical or gasoline energy. In other words, before going to a community sewer, buying a supermechanical contraption, or giving up altogether, try these tricks of the sanitary engineering trade.

Hillsides and Drainfield Construction

Slopes are, in themselves, not necessarily a bad place to distribute effluent. In some cases they can provide better drainage than a flat area. Many times the "daylighting" of wastewater from the drainfield occurs because the soils were too shallow or too clayey in the first place—not because of the slope. *Soils and slopes must be considered separately in judging a site for a drainfield.*

If the problem is slopes then follow these recommendations:

SERIAL LOADING IS A MUST FOR SLOPES. Many contractors try to use trenches stemming from a distribution box. Many times, as explained in chapter 6, all the effluent flows quickly to the bottom trenches and the top trenches wind up totally unused. With or without distribution boxes, flat-ground drainfields do not work well on slopes.

Serial loading can be accomplished by the "drop-box" method (see Public Health Service Manual) or by the use of plastic fittings. The latter is cheaper. Four plastic fittings—two 90-degree elbows, a 45-degree ell, and a tee—are attached so that the wastewater must "jump" from one infiltration trench to the next. This forces the first trench to fill before the second can be used. An area

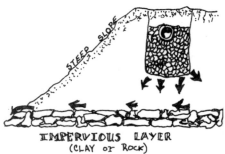

IMPERVIOUS LAYER
(CLAY or ROCK)

A slope may cause problems if the soils are too shallow to clean the water before it reappears on the surface or if the water causes erosion when it reappears (at the seep point). Deep soils on slopes (left section of top drawing) can be fine for slopes. On slopes of more than 20 per cent it is difficult to use a backhoe.

of undisturbed earth serves as a dam between the infiltration trenches and, together with the "jump," prevents wastewater from filling the lower trenches first. Caution: Do not dig into the undisturbed earth dam between trenches; dig only deep enough for the jump pipe.

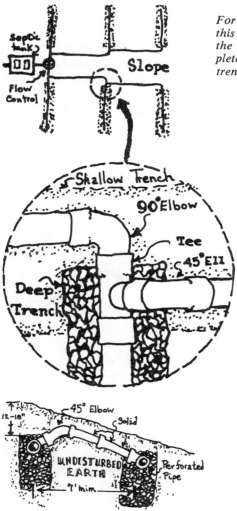

For serial (slope) layouts, this pipe arrangement forces the top trench to fill completely before the downhill trenches fill. The layout is shown schematically at left. Stippled areas are gravel trenches; other solid lines are solid pipe. Below left is a detail of pipe arrangement using a tee, a 90-degree elbow, and a 45-degree elbow.

TRENCHES SHOULD BE PARALLEL TO THE CONTOUR.

TRENCHES SHOULD BE AS LONG AS POSSIBLE. Many codes insist that trenches be only 100 feet long. This is a nonsensical rule, especially for drainfields on slopes.

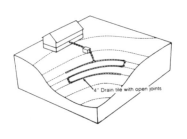

AVOID BUNCHING TRENCHES ON TOP OF EACH OTHER. For backhoe construction, the outer wall of a trench should be about seven feet from the nearest trench. This is a super-safety factor, not a hard-and-fast rule. (See Section V, "Making the Exact Map for Construction," in chapter 8.)

TRENCH BOTTOMS SHOULD BE ABOUT LEVEL. Off-grade trenches on slopes lead to unequal ponding and unequal use of trenches.

OFF-GRADE

ON-GRADE

NEVER BENCH A DRAINFIELD. ON STEEP SLOPES, USE HAND TOOLS. Slopes with a grade of 20 per cent (rising twenty feet in one hundred feet) cause most of the problems. Slopes of 20 per cent are about the steepest that a backhoe can work on safely. So humans, insistent on their machines, "bench" the slope (i.e., cut it into a series of steps) instead of using hand tools. This excavates the soils down to impermeable bedrock and leaves exposed vertical cuts where seepage can occur. Benching also compacts and smears the soil so much that the waste-water can no longer infiltrate and drain effectively.

BENCHED SLOPE

Soils with Hardpans and Seasonal Water Tables

When water perches on a hardpan or bedrock, it can flood the drainfield. If the drainfield is flooded for too long, it may turn anaerobic and clog. It is important to know the following: How close is the bottom of the drainfield to the groundwater? How long does the groundwater remain either in or right below the drainfield? Is the groundwater just sitting or is it moving through the drainfield? Is it fluctuating or permanent groundwater?

Three clues can help you decide how good or bad your groundwater situation is for a drainfield. Ideally, you can dig a small hole and watch the groundwater during the wet season. See how far up and down it fluctuates and for how long it is closest to the ground surface. This assumes you have a wet season available before you want to start building.

Another clue to the groundwater level is the soil colors and mixture of soil colors. If soils are saturated (permanent groundwater), the iron in the soil forms a gray compound (ferrous oxide). When the soils get good aeration ("insignificant" groundwater), the iron in the soil turns red-brown (ferric oxide, or hematite). Fluctuating groundwater produces a mixture of red-brown and gray colors. In addition, a third form of iron, called limonite, which is yellow in color, develops. Intermingling of gray, yellow, red, and brown indicates a seasonal groundwater table and the general depth of fluctuation. These *mottled* soils usually show good percolation in dry seasons but, come the wet seasons, the soils are easily saturated and may not drain the effluent.

Last, permanent or semipermanent groundwater tables nurture special water-loving plants. The presence of these plants is a warning that there may be wet-season troubles.

Soils that are continually wet from prolonged water tables or ponding nurture special plants. These plants, such as the rushes pictured (on the left) and the brambles (on the right) love water. Possible waterlogged soils can be spotted easily if you can recognize these plants. Others include poison hemlock, sedges, etc.

Curtain Drains

Drainfields in mottled or seasonally saturated soils can sometimes be made to work if artificially installed drains are provided. These are called "curtain drains" or "French drains" because French farmers have used them for centuries to control the wetness of their land. During rain storms, curtain drains effectively divert the surface runoff around the border of the drainfield and intercept some of the moving underground water and divert it from the field. There are many designs.

Areas of seasonal groundwater are very common in the West; they are best solved by on-site investigation of the soils and the

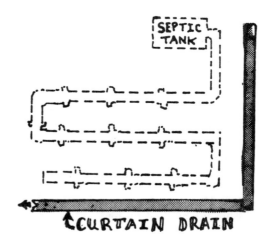

This curtain drain is catching subsurface water that moves down slope (right to left). The curtain moves the subsurface water around the drainfield. This helps prevent flooding. The curtain drain water goes somewhere like a ditch or stream.

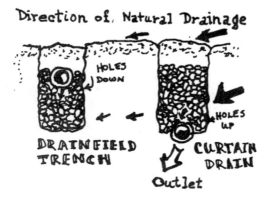

Note differences between a drainfield trench and a curtain-drain trench. Holes should be up to keep soil out of the pipe. For more detail, see page 127.

water table. In general, the main problem with *occasional* high groundwater is a slow-flush or sluggish drain in the house. Making sure the toilets and sinks are built well above the septic tank helps in this situation. A greater drop from the toilet and drains to the septic tank helps force the wastewater into the tank.

REMINDER: Mottled soils are usually found two to three feet below the surface of the ground. Colored soils much above or below this level could be indicative of something else—for example, humus or bedrock color.

REMINDER: Slopes can have seasonal groundwater (and mottling) just like flat ground. But draining effluent is not as difficult on slopes because the underground water moves downhill, diluting, purifying, and moving the effluent.

WARNING: Curtain drains will not solve severe water-table problems. If the groundwater completely floods the trenches, other techniques must be used (see below).

Community Drainage

Many subdivisions or communities must solve their high-water-table problems on a community basis. Culverts, ditches, surface runoff, and underground water movement of a complete area or watershed must be understood. Otherwise, one home dweller is just shunting his or her water into a neighbor's drainfield.

Shallow or Impermeable Soils and Long-Term Groundwater

Where the soils are too shallow or are very impermeable or are permanently saturated, septic-tank systems can be modified. Essentially, it is possible to *construct an artificial soil profile,* with appropriate plants on top, that will filter, drain, and evapotranspire the wastewater. Each system must be custom-designed for the soils, the slope, and the evapotranspiration rate of your region.

The commonest of these humanoid soil filters are the mound, the evapotranspiration bed, and the inverted sand filter. The shapes of these human-made beds of soil and the exact layering of sands and loams vary with the scientist.

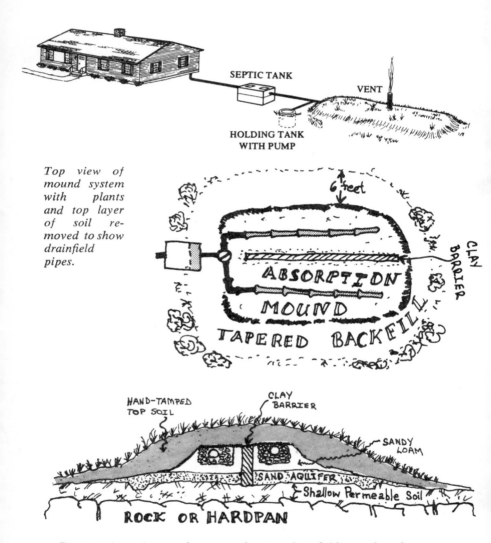

Top view of mound system with plants and top layer of soil removed to show drainfield pipes.

SEPTIC TANK

VENT

HOLDING TANK WITH PUMP

6 feet

CLAY BARRIER

ABSORPTION MOUND

TAPERED BACKFILL

HAND-TAMPED TOP SOIL

CLAY BARRIER

SANDY LOAM

SAND AQUIFER

Shallow Permeable Soil

ROCK OR HARDPAN

Cross-section of mound system showing drainfield trenches, human-made soil profile, and replaced topsoil.

Creating a Soil Profile

Making your own soil filter bed means trucking sand, gravel, and soil (usually sandy loam or loam) to your home site. The native ground surface beneath the mound is cleared of all plants and, if there is some soil, this should be rototilled and left rough. To insure clean water, at least two feet of loam or sandy loam should be laid down between the native ground surface and the bottom of the drainfield pipes. If loams are impossible to find or are too expensive, a series of finer and finer sands can be used.

The drainage pipes (pressurized or unpressurized) are laid in gravel above the two feet of soil filter. In the "mound" design, the filter bed and drainage pipes are covered with at least three feet of topsoil.

To increase the cleaning powers of the soil, all the engineering tricks to increase aeration should be used: dosing tanks, breezers, and alternate beds. In some mounds, especially those that use evapotranspiration, air vents are laid next to the drainage pipes. This brings air into the soil in a more reliable way—increasing microbial action and water absorption by plant roots.

An evapo-bed *designed by Bernhart. It is totally enclosed by plastic sheeting, with extra air brought in by small vent pipes. It uses only evapotranspiration to recycle wastewater. This design is further described in chapter 8 and Bernhart's excellent book (see Winneberger, Abney, Bernhart, and especially* Alternate Sewage Manual *in Bibliography).*

Moving the Clean Water

Moving the cleaned water away from the treatment site demands ingenuity of the designer. In areas where evapotranspiration
is possible, the mounds are usually larger and have pipes spreading the cleaned water among the plants. Some designers, preferring
not to use drainage pipes, use sand "aquifers" as conduits to
whiz the water about. These sand aquifers can be cheaper than
pipe, depending on the cost of pipe, and they also provide some
further treatment.

LUXURY SEPTIC TANK EQUIPMENT

Pumping Chamber with Dosing Pumps

With many mound systems, it is necessary to pump effluent after pre-
treatment in the septic tank. This assures even distribution of effluent
within the mound (stops puddling) and aids aeration. A special pump-
ing chamber is usually constructed between the septic tank and the
mound. These chambers have 500–750-gal. volume to store one day's
effluent plus a one-day emergency load. The pump should be selected
on performance curve (total head vs. capacity)—not horsepower.
Pumping distance must be kept short (less than 150 feet) to avoid
friction losses. Head is usually twice the elevation between pump and
drainfield pipes. Capac-
ity is 25 to 35 gallons
per minute, depending
on water usage. See
Franklin   Research
(4009 Linden Street,
Oakland, CA 94608)
for the "Cadillac" of
pumps   (the   Air-O-
Pump) and the Wis-
consin Alternate Sew-
age Manual for further
details.

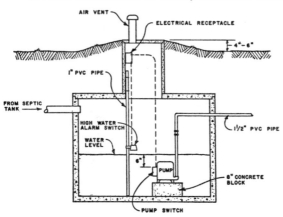

When evapotranspiration cannot be completely relied upon, the filter bed size is increased to insure that water will be cleaned. A curtain drain directs the cleaned water away from the filter bed. Either sand conduits or plastic pipes will work, and the choice depends on your home ecology.

The next chapter contains information on how to size your particular septic-tank system. These systems have been tried and used successfully all over the country. They are still cheaper than public sewers.

Nonelectric dosers (the Miller siphon) are available but tend to corrode within five years. The use of dosing tanks with regular drainfields is possible. But recent claims by dosing-tank manufacturers that this will allow a 50 per cent reduction in drainfield size have not been proven to our satisfaction.

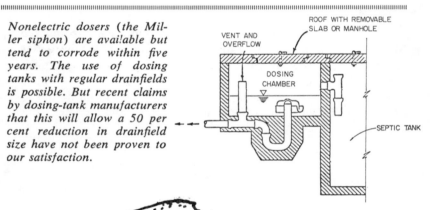

Even Drainfield Flow

In most cases, effluent must be pumped to the mound. To insure the wastewater can move evenly through the mound and won't puddle, pond, or daylight in one small section, Winneberger has made the following suggestion: Place the pressure pipe inside the regular drainpipe to protect the small holes of the pressure pipe from clogging. For example, the pressure pipe of 1¼" (with ⅛" holes every foot) is placed inside a 4"-diameter drainfield pipe.

Cold Climates

The northern states of the United States are all subject to frost penetration problems. The best guide to keeping drainfields working in deep cold is to follow local manuals and local advice. In general, little trouble occurs when the tank is at least one foot below the surface and the drainfield pipe is 18 to 24 inches below ground surface.

Flushing with warm water is said to help in winter months. Placing the tank below the frostline is usually not necessary, especially if you cover the septic-tank/drainfield area with straw. Freezing occurs most often when you don't use the system continuously.

WARNING: Sewer pipes under driveways and other hard-packed surfaces suffer most. This usually occurs because the insulating layer of snow is shoveled off. It's best to insulate sewer pipes under driveways.

Evergreens, such as cedar, continue to evapotranspirate during the winter. They can help drainfield function in cold climates.

Summary

Geology and water movement set the limits for drainfields. Major considerations are: (1) texture of the soil receiving the effluent; (2) thickness of the soil; (3) depth to groundwater and seasonal variation; (4) depth to bedrock or hardpan; (5) texture of bedrock (solid or fractured, with or without soil); (6) extent of bedrock or hardpan (where it directs effluent); (7) direction of surface runoff; (8) direction of subsurface runoff; (9) slope of land; (10) proximity to road cuts or embankments; and (11) proximity to natural watercourses. Any of these limitations, if extreme or not considered when designing the drainfield, may increase the chances of failure.

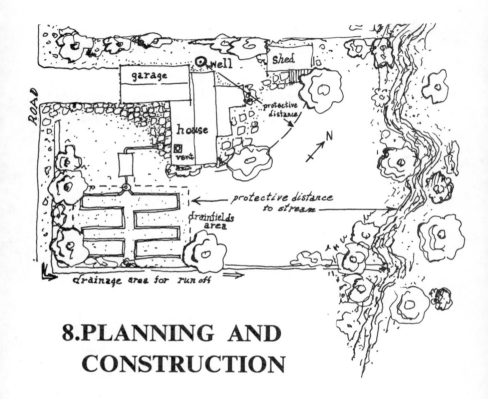

8. PLANNING AND CONSTRUCTION

Nothing in this chapter is foolproof. Home-site sewage treatment, using soils, is always filled with mystery and luck. You must balance costs, the space available, soils, climate, and dreams. *Local* know-how is crucial when designing your home-site system. In addition, every person has his prejudices. This chapter combines the *practical* experience of myself and Dr. Timothy Winneberger and my determination to make each aspect of septic-tank-system design as ecologically relevant and as cheap as possible.

The chapter discusses the design and installation of septic-tank/drainfield systems by stages: (I) Sketching the Home-Site Plan, (II) Decision Making, (III) Learning Your Home-Site Ecology, (IV) Calculating the Drainfield Size (V) Making the

Exact Map for Construction, (VI) Purchasing and Replacing Materials, and (VII) Construction Practices. It is meant to provide the most up-to-date knowledge and allow you to construct a home-site treatment system that will last as long as the house (fifty to seventy-five years).

I. Sketching the Home-Site Plan

On an outline map of your property (a scale of about one-eighth inch equaling one foot is easiest to draw on), sketch:
  a. The north arrow.
  b. The size, shape, and location of your home, outbuildings, proposed buildings, and hardened surfaces such as patios, decks, walkways, driveways, and swimming pools.
  c. The location of any wells, springs, creeks, or drainage channels on or near your land.
  d. The area you would like to remain green. This is the area of the drainfield, septic tank, and expansion area (if you need one).
  e. An eyed sketch of the contour lines, showing slopes, and maybe a side view if you think there will be grading or flattening.

WARNING: Slopes are special cases in home-site sewage treatment and must be carefully analyzed. Read chapter 7 before proceeding if you have a slope greater than about 5 per cent.

REMINDER: A southern exposure will help evapotranspiration and gardening.

This sketch will be the basis for the "hard" blueprint when you add county codes and drainfield size. Before you get too exact, a series of decisions have to be made.

II. Decision Making

Keeping in mind the amount of land you have, you must decide how much you can afford to spend on a sewage system, and how much you know about its climate, soils, and geology:

a. Do you want to have a greywater drainfield and a septic-tank drainfield? The greywater drainfield requires separate plumbing for the sinks, showers, and bath. Greywater drainfields use about half the amount of sidewall for the same amount of wastewater as black-water drainfields do.

b. Where will the surface runoff (rain) go? Directing the runoff from the roof or patios away from the drainfield area (in areas of heavy rainfall) and knowing where the runoff will leave the property line are both important.

c. What geological and climatic limitations might need further investigation or might restrict the possibilities? These include:
  1. A permanently high water table.
  2. A rise in the groundwater during the rainy season that might enter the bottom of the drainfield.
  3. The amount of rainfall.
  4. The amount of evapotranspiration.
  5. The depth to bedrock and the kind of bedrock.
  6. The existence of hardpans which are impermeable to water.
  7. Nearby watercourses and man-made cuts or embankments.
  8. The depth and texture and structure of the soil.

Much of this information will be discovered by doing a soil bore and asking neighbors about the water table, etc. This is explained in the next section.

REMINDER: Curtain drains (see chapter 6) are useful to lower seasonally high water tables.

d. What type of drainfield best suits your needs? As your understanding of climate, soils, costs, and re-use of water becomes more and more connected to your land, the kind of drainfield that suits your needs best will become increasingly obvious. If the soils are deep and have few geologic or water-table limitations, then *seepage pits* are usually preferred. If the soil or water limitations occur three to six feet below the ground surface, then *trenches* are usually preferred. If limitations are severe, then *shallow trenches* or a *seepage bed* must be used. In many areas, seepage beds can work with evapotranspiration only. These are called *evapo-beds*. Finally, if there is really no useful soil, then you must build an artificial soil profile (chapter 7) and/or switch to waterless toilets (chapter 2).

e. Do you want (1) a single drainfield? (2) a single drainfield with an expansion area? (3) a dual drainfield? or (4) a dual drainfield with an expansion area? All these kinds of drainfields (trenches, pits, or seepage beds) can be used continuously or in alternation. If you have only a small lot, a single drainfield is a necessity. If you have the space and can afford the cost, a dual drainfield will last longer and work better than a single one (chapter 6). In short, always choose a dual drainfield (vs. a single one with an expansion area) if possible. A dual drainfield with an expansion area is, of course, ideal, but you must have a large lot to accommodate it.

III. Learning Your Home-Site Ecology

You want to know what the highest groundwater level might be, whether the bedrock is fractured or solid, whether the bedrock cracks are filled with soil or have empty spaces, whether the soil permeabilities will allow water to infiltrate and how fast, which layers of soil are the best to use for infiltration, whether there is hardpan, etc.

Soils

The best way to get soil information is by asking honest and knowledgeable local people—the sanitarian, the contractor, your next-door neighbor, the local representative of the Soil Conservation Service, or the local engineering firm. You can get your own information by doing a soil bore, collecting soil samples, having the SCS analyze the soil texture for you, and by performing your own percolation test.

SOIL TEXTURE BY FEEL

The soil can be tested by feel in either a dry or wet condition. Moist soil can be tested more easily.

Sand: Individual grains are readily seen and felt. Squeezed in the hand when dry, soil will fall apart when the pressure is released. Squeezed when moist, it will form a cast that holds its shape when the pressure is released, but will crumble when touched.

Sandy Loam: Consists largely of sand, but has enough silt and clay present to give it a small amount of stability. Individual sand grains can be seen and felt readily. Squeezed in the hand when dry, this soil will fall apart when pressure is released. Squeezed when moist, it forms a cast that will not only hold its shape when the pressure is released, but will also withstand careful handling without breaking. The stability of the moist cast differentiates this soil from sand.

Loam: Consists of an even mixture of the different sizes of sand, silt, and clay. Crumbles easily when dry and has a slightly gritty, yet fairly smooth feel. Squeezed in the hand when dry, it forms a cast that withstands careful handling. The cast formed with moist soil can be handled freely without breaking. It is slightly plastic.

Silt Loam: Consists of a moderate amount of fine grades of sand, a small amount of clay, and a large quantity of silt particles. Lumps, in a dry, undisturbed state, appear quite cloddy, but they can be pulverized readily; the soil then feels soft and floury. When wet, silt loam runs together and puddles. Both dry and moist casts can be handled freely without breaking. When a ball of moist soil is pressed between thumb and finger, it will not press out into a smooth, unbroken ribbon but will have a broken appearance.

Clay Loam: A fine-textured soil that breaks into clods or lumps, which are hard when dry. When a ball of moist soil is pressed between the thumb and finger, it forms a thin ribbon that will break readily, barely sustaining its own weight. The moist soil is plastic and forms a cast that will withstand considerable handling.

Clay: A fine-textured soil that breaks into very hard clods or lumps when dry, and is plastic and usually sticky when wet. When a ball of moist soil is pressed between the thumb and finger, it will form a long ribbon.

WARNING: The Soil Conservation Service has general classifications of soils, but these are *general* and may not apply to your own local lot. Only by taking soil samples (with depth indicated) to the SCS can you find out if your soils are typical. The SCS usually will analyze soil samples free of charge. Some pleasant members of SCS will even help you take the soil bore, collect samples, and analyze them right on your home land. Remember, even if the official SCS classification says, "very severe for drainfield use," your home-site soils could be unique and good for sewage treatment.

The *soil bore* should be about ten feet deep. If you encounter groundwater, record the depth. If you encounter bedrock, try to collect some and see if there is soil in the crevices or not. If you encounter a hardpan, see if you can dig deeper. Sometimes, you can perforate the hardpan and find soils below it that infiltrate well.

REMINDER: Look for mottling of all layers to see if there may be a seasonal groundwater problem (chapter 7).

REMINDER: You must know whether the soils are predominantly clay, silt, or sand. Do not guess. A sieve analysis, used by the Department of Agriculture, is the most economical way to analyze soil. You do not want to be slipshod about the soil texture or your guess at drainfield size will be inaccurate.

Percolation Test

The percolation test should be done in the soil that will receive the effluent. If there are many different layers, then a percolation test should be done in each.

WARNING: Some seepage pits in desert areas are thirty to forty feet deep. Don't do a perc test in these holes. Try to extrapolate permeability from the soil sample (see Abney's chart). It's hard to climb out of a collapsing hole.

REMINDER: The percolation test is most accurately performed in the wettest season.

Bernhart's Wastewater Loading Rates for Drainfields

Percolation Rate Min./in.		Maximum Loading Rate gal./sq. ft./day
Sand	1	0.63
medium sand	5	0.56
fine sand	10	0.47
fine sand—silt	15	0.39
silt—sand	20	0.32
silt—loam—sand	30	0.23
loam—silt	45	0.12
loam—clay—silt	60	0.05

The loading rates are for totally anaerobic conditions.

Abney's Loading Rates

Soil Texture	Sidewall for Each Gallon
Clay/silt (less than 0.1 mm)	4 sq. ft.
Sandy soils (greater than 0.11 mm; less than 1.0 mm)	2.5 sq. ft.
Coarse sands,* small gravel (greater than 1.0 mm)	1.5 to 2 sq. ft.

* When soils are this coarse, check drainage to make sure no nutrients or pathogens can reach them.

Ryon's Corrected Criteria, Maximum Wastewater Loading Rates for Drainfields

Percolation Rate Min./in.	Maximum Loading Rate gal./sq. ft./day
1 or less	2.5
2	2.0
3	1.6
4	1.4
5	1.3
10	1.0
20	0.72
30	0.48
40	0.42
60	0.36

* Active infiltration surfaces are sidewalls of disposal fields.

REMINDER: Perform at least one test for each layer to be used. The wettest season of the year is the best time to make the test.

The Test Hole

a. Dig or bore a hole with handtools 13 to 14 inches in diameter.

b. Remove any smeared surfaces from the sides of the hole to provide as natural a soil interface as practical to infiltrating waters. Remove loose material from the bottom of the hole and add an inch or two of coarse sand or fine gravel to prevent the bottom from scouring.

c. Presoak the hole carefully, never filling it deeper than about 8 inches with clean water. Do not pour the water into the hole from much distance; ease it in gently. If you know that the soil has low shrink-swell potential and its clay content is not high (perhaps less than about 15 per cent), proceed with the test. If not, let the hole rest overnight.

Measuring the Stabilized Percolation Rate

a. Fill the hole with clean water to exactly 6 inches above the soil bottom of the hole (do not consider a layer of protective gravel as the bottom of the hole). With a float gauge and a timepiece, determine the time it takes the water to recede exactly 1 inch. Refill and repeat the process until successive time intervals needed for 1 inch to be absorbed indicate that a stabilized rate has been obtained. Do not allow much time to pass between refillings.

b. Report the stabilized percolation rate in minutes per inch.

Results

If the percolation rate is more than 60 min./inch, then the soil is of questionable value as a disposal field. In these questionable areas, a minimum of three to five percolation tests should be run. The tests should be distributed over the area proposed for the drainfield and future expansion of the system. If these percolation tests indicate all the soil area to be of questionable value, then alternate means of waste disposal must be considered.

PROCEDURE

A smooth metal rod, ½ inch in diameter, and about 18 inches long, has a sharpened end which is driven into the ground beside the percolation test hole. With experience, one can position the rod to best advantage.

The graduated, transparent plastic tube is about ½ inch in diameter and 14 inches long. Graduations are provided by a plastic tape, printed in inches, graduated into tenths. (Scalafix is one brand sold through chemical-supply houses.)

A solid plastic rod about ⅛ inch in diameter slides into the graduated tube and the bottom end is glued into a spherical plastic float, about 2 inches in diameter. The length of the rod is about 18 inches, but should be adjusted so that enough is inside the graduated tube to hold the rod as rigidly in place as practical. The gauge is left in place during the course of the test; the graduated tube should not be moved. The graduations permit interpolation to within about 0.02 inch. This accuracy is more than adequate for the test.

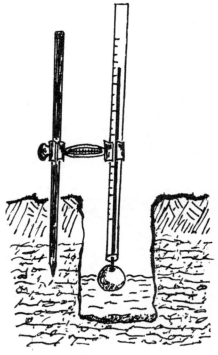

A universal clamp fits onto the rod. This clamp, available from chemical-supply houses, twists conveniently in different directions. It has a thumb-screw at each end; it can clamp onto the rod and at the same time hold the graduated plastic tube vertically over the test hole.

—*From* Current and Recommended Practices for Subsurface Waste Water Disposal Systems in Arizona *by John Timothy Winne-berger, Ph.D., and John W. Klock, Ph.D.*

DEEP HOLES

To make a perc test in deep soils, dig a larger hole down to the layer you will be using to infiltrate effluent. Make the perc test in a smaller hole within the large hole as pictured above. Don't attempt perc tests in holes more than 10 feet deep; you might wind up buried alive.

COLLAPSIBLE SOILS

If soils tend to collapse, place a perforated pipe vertically in the hole and carefully pack gravel or some supporting material between the pipe and the hole wall. Perform the test within the vertical pipe and adjust calculations to account for the displacement of water by the gravel or whatever is used to support the sides of the hole.

NOTE: Without taking into account the gravel in the hole, you will get a faster rate of drop than actual water absorption. For each 1-inch drop, only 60 per cent was actually water; 40 per cent was gravel space. For example, a 2-inch drop in water level is really only a 1.6-inch drop. Remember to adjust the drop by accounting for gravel.

IV. Calculating the Drainfield Size

Step 1: Total Water Use

If you have no luxury appliances and take care not to waste water, you should estimate the use of 55 gallons per person per day for your household. If you have luxury appliances (dishwasher, clothes washer, garbage grinder), then estimate 75 to 100 gpppd. If you have lots of negligent people in your house who have extravagant water habits they just can't change, then estimate 150 gpppd. If you think you need more than that, you are greedy.

Step 2: How Much Will Go into the Septic Tank?

Not all the wastewater must go into the septic tank. The greywater can be separated from the black water. The greywater (or part of it) could go directly into a seepage pit or subirrigation system. For instance, if a family wastes 200 gallons a day but only the toilet goes into the septic tank, then only 100 gallons actually enter the septic tank each day.

Step 3: Loading Rate

The loading rate is the amount of wastewater that can infiltrate into one square foot of your drainfield soils each day. The loading rate can be determined from the percolation test and/or the soil-texture test, modified by information from neighbors, contractors, local sanitarians, and the Soil Conservation Service reports and comments. The charts on pages 81–82 will be useful in translating the perc test and the soil-texture test into loading rates.

Step 4: The Amount of Sidewall

To determine the amount of sidewall needed, you simply multiply the amount of water going into the septic tank by the loading rate. For example, if 200 gallons per day go into the septic tank and your clayey soils require 4 square feet of sidewall for each gallon of wastewater, the amount of sidewall required is: $200 \times 4 = 800$ sq. ft.

WARNING: Never use the bottom area in calculating the amount of infiltrative surface. *Use sidewalls only.*

WARNING: Do *not* use the "number of bedrooms" or "number of persons" to determine sidewall area, as prescribed in many codes. Only use estimates of actual water use and the perc test and/or soil-texture test.

Step 5: Translating Required Sidewall Area into the Amount of Trench or Seepage Pit

Using all the geology, soil, and climate information available, you must now translate the amount of sidewall needed into the trench, pit, or seepage bed shape.

─────────────── SIDEWALL COMMENTS ───────────────

Two Different Soil Layers

Where there are two different soil layers, you will want to use the most permeable. With two different layers, calculating the amount of sidewall is more complicated.

a. Estimate the proportion of each soil layer to be used for infiltration. Example: The sandy layer may be ⅓ of the total infiltration surface and the clay layer ⅔ of the infiltration surface.

b. Determine separately the amount of sidewall area to be used for each layer. Multiply the fraction times the loading rate times the amount of waste-to-the-septic-tank. Example: ⅔ (proportion of the sidewall in the clay layer) × 4 sq. ft. (clay-layer loading rate) × amount of wastewater.

c. Add up all the sidewall areas for all the layers utilized. Go to *Step 5.*

The Conservative Approach

If your neighbor says that his/her drainfield works all right and you don't want to pay for perc tests, you can be very conservative. This conservative approach says, if water percs at all, then use ¼ gallon per sq. ft. (1 gallon for each four sq. ft.) of sidewall. This is equivalent to saying all your soils are clayey.

You must still look at geologic limitations to see how deep the trench or pit must go. But you can avoid certain county codes requiring very expensive perc tests. This conservative approach results in more sidewall, a bigger drainfield and higher costs. You must decide.

A. Trench

The effective trench depth is the distance from the top of the pipe down. For instance, if the trench is five feet deep and the top of the pipe is one foot below the level of the soil, the *effective* depth is only four feet.

$$\text{Running feet of trench needed} = \frac{\text{total sidewall required}}{2 \times \text{effective trench depth}}$$

(The "2" in the bottom of the formula is because each linear foot of trench has two sides and the effluent will, of course, infiltrate both sides.)

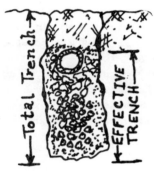

Example: If 1,600 square feet of sidewall area are needed and the effective trench depth is four feet, then

$$\frac{1,600}{2 \times 4} = 200 \text{ linear feet of trench}$$

B. Seepage Pit

The effective depth of a seepage pit also starts from the top of the pipe down. Seepage pits are commonly 3 to 8 feet in diameter

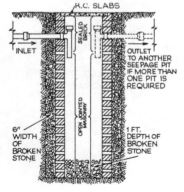

and 7 to 16 feet deep. Some are much deeper. They are a definite safety hazard as the walls may collapse. It may be too dangerous to perform perc tests if you must climb into the pit. To calculate the number of seepage pits you need, first determine the diameter and depth of the hole you can excavate. Second, determine the effective sidewall area in this pit. Third, divide area needed by the effective sidewall area of your pit. This will give you the number of seepage pits required.

Example: You decide you can dig seepage pits with a 4-foot diameter and 15 feet deep, giving an effective depth of 14 feet. The sidewall surface area is:

$$\pi \times \text{Diameter} \times \text{Height} = \frac{22}{7} \times 4 \times 14 = 176 \text{ square feet}$$

If you need a total of 1,600 square feet of sidewall, then you will need nine seepage pits of this size.

For construction details see Appendix 1.

C. *Seepage Bed*

Seepage beds are usually no more than 18 inches deep. They use both infiltration and evapotranspiration to recycle wastewater into the soil and atmosphere. Bernhart has developed an empirical chart based on soil texture and assuming 0.12 gallons will evaporate each day per square foot of seepage bed. This evapotranspiration data is for southern Ontario and will increase as you go south into the United States. It assumes 200 gallons a day from the household—no more. It also assumes the seepage bed is well ventilated, well planted, and well crowned to remove most rain.

Some seepage beds are lined with plastic, so that effluent will not escape. Bernhart estimates that 1,700 square feet of evapobed will work just about anywhere in the U.S.A. This is a 40-by-45-foot piece of land set aside only for *permanent* greenery and *aerobic* seepage-bed construction. Details change as you move south or to areas of high rainfall. I refer you to his book (see Bibliography).

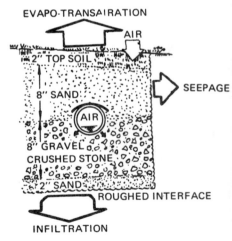

V. Making the Exact Map for Construction

We must now redo the original sketch with "hard" numbers. Starting from the property line is usually easiest. First, draw in all the setbacks either required by the county or needed for well protection, etc. Next, draw the outlines of the drainfield, dual drainfield, greywater drainfield, and/or replacement areas. Always draw within the setback borders. Follow these guidelines:

a. Serial distribution within the drainfield is definitely preferred. On slopes, it is a must. Special designs are needed in difficult terrain (see pages 90–93).

b. Where linear trenches are used, each trench can be as long as possible but should be separated from the next trench by a nonperforated overflow pipe in *undisturbed* earth. (If the ground slopes, see chapter 7.)

c. The drainfield trenches or seepage pits should be spaced apart. Measure from the outer well of one drainfield unit to the outer wall of the next nearest. This space should be 1½ times the effective depth (depth below the pipe).
   Example: A trench with a 4-foot effective depth should be about 6 feet from the nearest outer wall of the next nearest trench: $1\frac{1}{2} \times 4 = 6$.

d. If you are using a curtain drain around the drainfield (chapter 7), make sure you have allowed room for it. Care must be taken, when installing curtain drains, to avoid contamination from the drainfield. Digging the drain about 10 feet away from the nearest trench or pit is usually more than adequate.

CONSERVATIVE SETBACK REQUIREMENTS

Site Features	Setback to Septic Tank	Setback to Drainfield
Buildings	5 feet	5–10 feet
Adjoining property lines	5 feet	5 feet
Wells (on-site or neighboring)	50–200 feet	50–200 feet
Natural watercourses	25 feet	50–100 feet
Cuts or embankments	25 feet	50–75 feet
Swimming pools	10 feet	25 feet
Water lines	10 feet	10 feet
Walks and drives	5 feet	5 feet

All these requirements are very conservative for clayey soils but could be lenient for shallow wells in medium to coarse sand.

Well siting depends on the depth of the well as well as soils. Bernhart is a useful reference. For example, he states that a 90-foot horizontal distance from well to drainfield is enough when the well is:

140 feet deep and soils are medium sand
120 feet deep and the soils are fine sand and silt
100 feet deep and the soils are silt, loam, and sand
80 feet deep and the soils are loam and silt
60 feet deep and the soils are loam, clay, and silt
40 feet deep and the soils are clay and silt

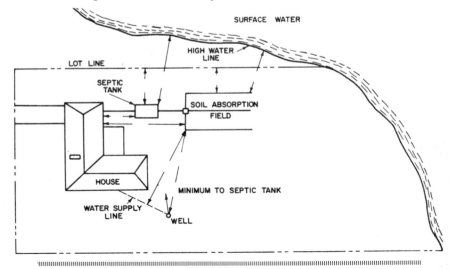

A DUAL DRAINFIED USING TRENCHES

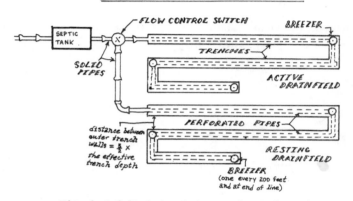

This drainfield design irrigates plants above it every other year. To irrigate plants each year, use an interlaced design.

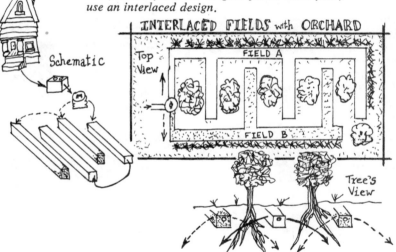

If you are using the dual drainfield system, you can still irrigate the same trees or plants each year by interlacing the trenches. The diagrams show how this is accomplished. Note: A dual drainfie'd without interlacing will supply water to plants on alternating years. This may kill plants.

VI. Purchasing and Replacing Materials

Diversion Valves

These flow switches should be of durable, acid-resistant materials. You can make your own.

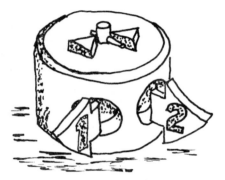

Hancor, Inc. (Findlay, OH 45840), makes a flow-control switch and drainfield pipes in plastic. Franklin Research (4009 Linden St., Oakland, CA 94608) also makes a flow-control switch (left) plus many other septic-tank/ drainfield control devices such as level-sensor, effluent pumps, etc.

Homemade flow-control switch using a concrete box and plastic ell. The turned-up ell stops effluent from going into the drainfield on the left. Next year the ell will be inserted on the right. Pipes and ell are all the same diameter (usually 4 inches). A rubber ball can be inserted in the pipe instead of the ell. The concrete box should be made of portland cement (see Appendix).

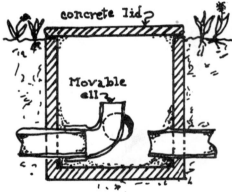

Septic Tank Materials, Size, and Design

a. The tank itself should be watertight. Durable, noncorrodible materials like reinforced and precast concrete or fiberglass should be used.

b. The tanks should be coated with bitumastic tar. This will retard corrosion.

   c. Minimum tank size should be about 800 gallons for a single-family dwelling. Remember, the bigger, the better—if the cost increase is not prohibitive. For two- or three-family dwellings, the Public Health Service *Manual* is useful. A 1,200-gallon tank is the biggest you'll need for a single family.

   d. A tank with compartments is much better than a plain box. The dividers can be redwood, treated plywood, properly prepared and treated concrete, or concrete block. The best design is the three-compartment meander design. Next best is the three-compartment flow-through design. Next, the two-compartment flow-through.

SEPTIC TANK MATERIALS

REDWOOD
Lifespan: 30 years. Rots more quickly at waterline or from outside in. Should be tarred on inside. Costs vary.

FIBERGLASS
Lifespan: 30 years (?). Most fiberglass tanks are too thin. Water in surrounding soils causes buckling. Some float. Buy or use fiberglass standards in the Uniform Building Codes. Make sure you can add compartments. Expensive.

PLASTIC
Lifespan: 30 years (?). Not yet manufactured. Believed to be a good re-use of non-biodegradable plastics.

PREFORMED CONCRETE
Lifespan: 20 years. Should be properly cured. Walls should be reinforced. Portland cement useful on side (2¼"). Bitumastic tar inside.

TILE CYLINDERS
Lifespan: 20 years (?). Usually terra-cotta. Used in a series.

METAL
Lifespan: 7 years. Corrodes easily. Coated metal only—if you must.

CONCRETE BLOCK
Lifespan: 20 years. Use heavyweight blocks to prevent collapsing. Hard-burned brick is a good material. Bitumastic tar inside.

Instructions for Preparation of Water-tight Cement Given in Appendix.

e. Rectangular tanks should be at least four feet deep below the waterline. A series of connected cylinders (page 71) will serve just as well.

f. Inlet and outlet ells (or *capped* tees) should be used to prevent solids from backing up. If plastic tees and ells are cheap and available, they serve the purpose well. Concrete designs are also fine.

WARNING: If tees are used, cap them with screen. Recent studies have shown that insects live in the septic tank and leave by flying through the top of the "tee" and through the vent from the house.

g. The tank should be easily accessible to permit maintenance. Each tank should have two covered manholes, allowing entrance to the different compartments *and* easy cleaning of the baffles. Covers should be watertight and, if you want added ease of access, protected by risers as shown in the diagram.

h. The inlet pipe from the house should be at least 4 inches in diameter. The outlet pipe should be the same diameter as the drainfield pipe.

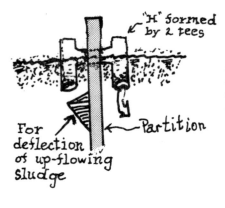

"H" formed by 2 tees

For deflection of up-flowing sludge

Partition

Tees can be made of cast iron, vitrified clay, concrete or plastic. They should be chosen to match at least the life span of the septic tank materials. Between compartments, two tees can be attached to form an H. This is the most effective baffle to prevent suspended solids or grease from flowing from one compartment to the next.

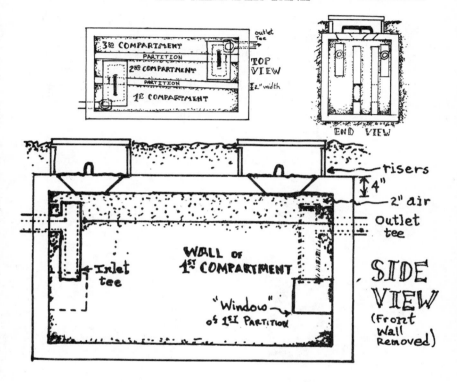

A 1,200-gallon tank might be 4 feet wide, 4½ feet deep, and 10 feet long.

—Partitions do not have to be watertight.

—Wastewater depth should be at least 4 feet.

—Width, at least 4 feet in rectangular tanks.

—Tank top not more than 2 feet below ground (*probably less*.)

—If a tee is used for inlet and outlet, then it should reach 18 inches below wastewater and as high as practical (6 inches is plenty) above the waterline. Top of tee should be no closer than 2 inches to tank top.

—First compartment should have the largest volume (40 to 50 per cent); second compartment should be larger than third (30 to 40 per cent)—in meander design.

—Windows can be replaced by double tees.

—Outlet tee must be 1 to 3 inches below the inlet.

Drainfield Materials

a. Gravity-flow pipes should have a minimum diameter of 3 inches. A greater diameter is all right as long as the pipes are not fragile or a safety hazard. Pressure pipes (which handle pumped effluent from dosing tanks or pumps) need only be as big as the outlet pipe from the pump. (Holes in pressure pipes can be hand-drilled.) All the pipes should be of strong and durable material like vitrified clay and certain plastics.

WARNING: Bituminous-paper pipes (like those produced by Orangeburg) will uncoil and rot very fast—don't use them.

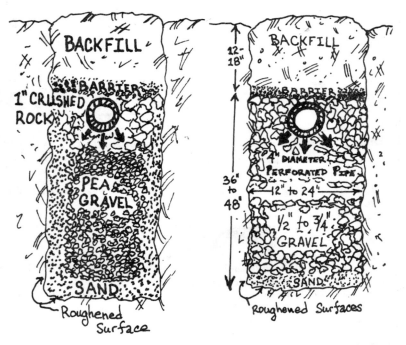

IDEAL TRENCH FILL
(*Expensive and more difficult*)

LAYERED TRENCH FILL
(*Cheap and easy version*)

b. The ideal fill for the drainfield trenches should grade evenly from large stones near the drainfield pipe to sand near the sidewall. Large stones lying next to the infiltrative surface can stop infiltration at the point of contact. To minimize soil-texture differences between the fill and the natural soil, use a column of pea gravel in the center of the excavation, flanked by two columns of sand. Filling trenches this way is a time-consuming chore and the fill is more expensive, but it's better for infiltration. (See diagram for instructions.)

Drawing on left shows how infiltration is reduced when large stones are placed next to soil. Drawing on right shows how a gradual change from gravel to sand to soil makes for ideal infiltration.

c. For a good standard trench (with no sand or pea gravel and no layering), use the smallest gravel available. The best is one-half inch to three-quarter inch. Gravel larger than three-quarter inch will reduce infiltration significantly.

d. Materials for the barrier between the trench fill and the backfill over the pipe are listed in the next stage.

VII. Construction Practices

WARNING: If construction is on slope, see chapter 7.

Septic Tank

a. The septic tank should be located so that it drains the household plumbing easily by gravity. No sink, bath, or toilet should be lower than the septic tank.

b. Because of the possibility of leakage, the septic tank should be set away from the building foundations (five feet is plenty).

WARNING: A septic tank may sink and bend or break its pipe connections. Be sure the septic tank rests on solid ground. Allow time for it to settle before connecting pipes. Water the hole to help compaction.

WARNING: If the tank is prefabricated and installed by a contractor, make sure it's not put in backwards.

c. The top of the septic tank should not be overloaded with backfill; one foot of earth on top is plenty. Don't dig the hole deeper than necessary.

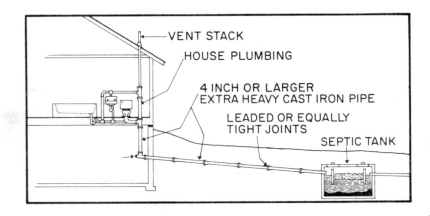

d. Risers make access to the pipe very easy. Install them if possible (see diagram page 121).

e. Remember to vent pipe(s) from the house.

Drainfield Construction

WARNING: Trenches should be dug with a backhoe or with hand tools. Seepage pits can be dug with a backhoe, bucket auger, or hand tools. Do not use spiral augers or trenchers, which will compact the sidewalls and smear the infiltration surface—ruining permeability.

WARNING: Do all drainfield digging in the dry season and/or cover trenches if there is no dry season. Wetting the exposed infiltration surface may clog soils before they are even used.

REMINDER: Use flat ground distribution only on flat ground. Never use a distribution layout designed for flat ground on slopes. Serial distribution is always preferred.

a. The minimum linear-trench width should be not less than one foot.

b. The ideal trench fill can be installed by the use of removable plywood sheets, as shown in diagram.

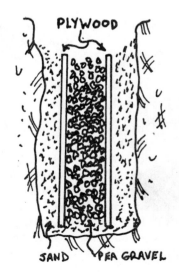

c. The tops of the drainfield pipes must be lower than the tops of the septic-tank pipes.

WARNING: Water in the drainfield can back up into the tank. Drainfields that are higher than septic tanks cause many problems.

d. A barrier layer is needed between the gravel fill and the soil cover. Any permeable material (such as straw) that will form a solid barrier is acceptable; the permeability allows evapotranspiration. The ideal barrier is a thin layer of roofing gravel, then a thin layer of coarse sand, and finally a thin layer of fine sand.

WARNING: Covering the trench fill with tar paper, plastic sheets, or other impermeable materials will prevent upward movement of water to plants and to the soil surface. Don't use such materials.

e. Drainfield pipe should be laid just about level. Little variations (1 to 2 per cent) don't matter. If you can do it by eye, that's fine. A casual use of the level helps too.

SEEPAGE-PIT WARNING: If the pit is more than ten feet deep, it should be filled with small stones for safety's sake. It's hard to climb out of a thirty-foot seepage pit.

Curtain Drains

Curtain drains should be of the same material and diameter as the drainfield pipe. The curtain drain should be as narrow as feasible. Wider helps not at all.

If the curtain drain is to remove perched groundwater, then the pipes of the drain should be laid into the top few inches of the hardpan. The perched water sitting on the impermeable layer will be drained away faster.

Curtain drains should be about 10 feet from the nearest drainfield trench. When pipe is used, the curtain-drain trench must be as deep as the drainfield trench. If no pipes are used (the trench is filled with sand), then the curtain drain should be about 2 feet deeper than the drainfield trench. Curtain drains move water by gravity away from the trenches. The grade should be 5 per cent. (See page 94.)

The outlet is usually a stream or a street drain. Make sure you have an outlet that can take added water.

WARNING: A common mistake of contractors is connecting drainfield pipe to the curtain-drain pipes. Be careful.

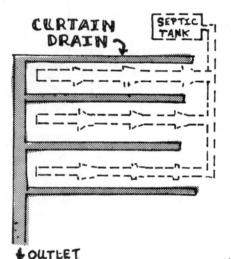

CURTAIN DRAIN
SEPTIC TANK

Curtain drain diverts subsurface water away from drainfield. In drawing at left, subsurface water is moving from the top to bottom of page.

OUTLET

PERMEABLE SOIL

HOLES UP

To Gravity Outlet

HARDPAN

The holes are up to keep soil out of pipe. If you put gravel under curtain-drain pipe then the direction of the holes doesn't matter.

Breezers

A minimum of one breezer for every 200 linear feet of distribution pipe *and* one breezer at the end of the distribution pipe are adequate. Seepage pits should have one breezer for each pit—if the pit is wide, two are good.

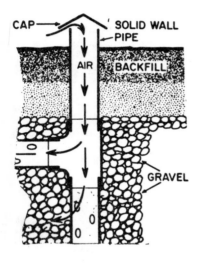

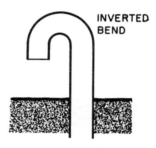

Location of breezers can be found on page 117. Breezers are either inverted bend pipes or stove pipes connected by a tee to drainfield pipe.

9. MAINTENANCE AND HOME CARE

NO DIGGING NECESSARY

WILL CUT THROUGH

ROOTS RAGS
GREASE
OR ANY TOUGH
HARDENED
ACCUMULATED
OBSTRUCTION

Most readers will not have the luxury of installing and guiding the construction of their own septic-tank system. Instead, they will be saddled with a former contractor's work, details of which may be unknown and long forgotten. They also do not know what the habits and water use of the previous homeowners were. Was the septic-tank system used all the time? Maintained? Abused? Neglected?

Know Your System

Many homeowners may not even know they have a septic-tank system, especially when they move from the city or just have a weekend home. The toilet flushes, and that's all they care about. This ignorance is bound to backfire and be costly. After reading this book, you should make a map of your septic-tank system and have it readily available. For instance, if you pay the man who pumps your tank by the hour.

It can take two to three hours just to find some septic tanks because the resident doesn't have the slightest idea where to look. Many new homeowners cement a driveway or patio right over the drainfield—totally eliminating aeration and plant life. Knowing the history and location of your septic-tank system will enable you to predict when the tank needs pumping, what the cause of any problems could be, how to plant your garden, and where to landscape.

Water Conservation

We have stated again and again that the less water used, the longer the lifespan of the drainfield and the better the septic-tank system will work. Chapter 4 lists all the do's and don'ts.

Chemicals and Microbes

In chapter 5 we emphasized that you must treat microbes well if you want them to digest sewage thoroughly. Small amounts of soaps, detergents, bleaches, drain cleaners, or lye will not harm large septic tanks, but an overdose can be fatal to your bacteria.

Waste brines from household water-softeners will change the clays in your drainfield. Never use them (or else, have a special drain), as they result in soil clogging.

Baking soda is one of the few additives that may actually help septic tanks. Baking soda helps keep the pH (the acidity or alkalinity of the septic tank) from fluctuating. The makers of Arm and Hammer estimate that one cup per week is enough to keep the septic tank buffered. Adding baking soda will also help get rid of odors.

For other do's and don'ts, see chapter 5.

Rainwater Runoff

A common mistake in home maintenance is letting water run onto the drainfield. All gutters should direct water away from the drainfield. All roofs that shed rain toward the drainfield should have gutters to prevent this. Patios and driveways should be located far from the drainfield or be ditched so that water from their surfaces will flow in another direction.

Pumping and Sludge

The septic tank can store only so much sludge and scum. After that, the sludge and scum begin to flow into the drainfield, clogging the soils. If this condition is allowed to persist, the toilet won't flush and the sinks won't drain. If this occurs, you have waited too long . . . damage to soils is underway. *Every two or three years the septic tank should be opened, the level of the sludge and scum checked, and the tank emptied if need be.*

The most difficult part of checking on the sludge and scum in your septic tank is opening the tank. Make sure you know where the tank is and install risers (see page 121) if you don't already have them. *Check the second (or third) chamber first.* If there is no sludge or scum in the last chamber, then that's all you need to know. If there is sludge in the last chamber and it is close to the outlet, then you should pump the tank. If you are unsure, check the first chamber(s). If they are chock full, then maybe you won't want to wait another year.

To determine the *scum* level, gently break the scum till a clear space can be seen. This should reveal how thick the scum is. If there's lots of scum in the first chamber but none in the last chamber, then scum is not the problem. If the scum in the last chamber looks as if it's about to go into the outlet pipe (within three inches), then it is best to pump.

To determine the *sludge* level, use a plastic tube. Remove scum with a shovel if it is too thick to penetrate. If the sludge is too thick to penetrate, churn your tank with the shovel. Otherwise, just insert the tube down through the scum and water into the sludge. The tube will give you a cross section of the tank.

In most communities, there is a nearby septic tank pumper. You

Scum

Sludge

To see if you need to pump, use a clear plastic tube about five feet long. Its diameter should be about the thickness of your thumb—around one-half inch. Carefully lower the tube all the way to the bottom of the tank. Then cover the top hole with your thumb (like you used to do as a kid with drinking straws), remove the tube carefully, and have a friend wipe the tube off. You should be able to see the three levels: scum, "clear" wastewater, and sludge. The federal government suggests using a terry-cloth towel attached to the bottom of a long stick to determine sludge depth. We have never known this to work.

call the truck and it arrives and sucks all the sludge, scum, and clarified liquid into a tank. *Don't ever wash your tank.* The remaining liquid contains enough bacteria to start a new colony of digesters.

If no truck service is available or you can't afford it, then you must pump the tank out yourself. This is illegal in some places because of the health risks. As we stated earlier, if no person with a water-borne disease has used the toilet recently, then it is impossible to have the disease living in your tank; but take precautions like wearing gloves and keeping children away. Dig a pit about two feet deep and five feet square away from the house and daily household traffic and buy a bale of straw, or get your own leaves or some other mulch. Using buckets, transfer the liquid and the scum from the tank to the pit. Compost the sludge (alternate layers of sludge and layers of straw), then cover it with about a foot of dirt and, if children are around, a plastic tarpaulin. Leave until composted thoroughly (six months to a year).

If your tank is commercially pumped, make sure the pumper is not emptying your sludge into a river or a storm sewer. *That's gross pollution.* The pumper should either take the sludge to a sewage plant (and hope it doesn't pollute too badly) or have a pit similar to the pit you would dig at home.

The price of pumping has sky-rocketed in many areas of the country. In parts of California, a single pumping may cost $80 to $90. Don't pump unless you need to. The Public Health Service has studied scum and sludge accumulation. For a family of four, a tank accumulates 98 gallons of scum and sludge by the end of the first year; 188 gallons by the end of four years. For an 800-gallon tank, this means that less than one quarter of the tank (22.5 per cent) is being used for solids storage after four years of service. Even after eight years, 60 per cent of the tank space will be free for the wastewater to pass through. Larger tanks store even more. In short, if you are not overloading your tank's capacity, it should need to be inspected less than once every two years. Some tanks go for twenty years without needing pumping, but, until you know your tank (and its bacterial abilities), check it every two or three years.

Septic Tank Problems and Cures

The symptoms of septic tank problems are very few: foul odors, poor flushing or draining of sinks, and evidence of erosion nearby. But these few symptoms may have innumerable causes. Good persistent detective work is needed to trace the symptom to the cause. The "Detective Chart" gives a pretty complete series of steps to determine the cause of any septic tank problem.

If you did not install the septic-tank system yourself, re-member:

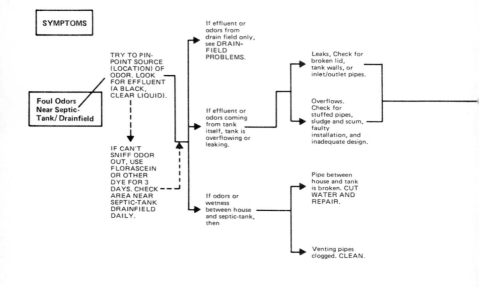

SYMPTOMS

Foul Odors Near Septic-Tank/Drainfield

TRY TO PIN-POINT SOURCE (LOCATION) OF ODOR. LOOK FOR EFFLUENT (A BLACK, CLEAR LIQUID).

IF CAN'T SNIFF ODOR OUT, USE FLORASCEIN OR OTHER DYE FOR 3 DAYS. CHECK AREA NEAR SEPTIC-TANK DRAINFIELD DAILY.

If effluent or odors from drain field only, see DRAIN-FIELD PROBLEMS.

If effluent or odors coming from tank itself, tank is overflowing or leaking.

Leaks, Check for broken lid, tank walls, or inlet/outlet pipes.

Overflows. Check for stuffed pipes, sludge and scum, faulty installation, and inadequate design.

If odors or wetness between house and septic-tank, then

Pipe between house and tank is broken. CUT WATER AND REPAIR.

Venting pipes clogged. CLEAN.

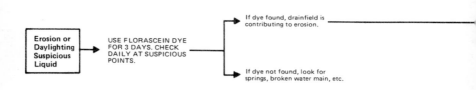

Erosion or Daylighting Suspicious Liquid

USE FLORASCEIN DYE FOR 3 DAYS. CHECK DAILY AT SUSPICIOUS POINTS.

If dye found, drainfield is contributing to erosion.

If dye not found, look for springs, broken water main, etc.

Poor Flush or Poor Drain

SNAKE TOILET OR PUT DRANO IN DRAINS

If still flushes poorly, OPEN TANK.

Check sludge and scum, design or faulty installation.

If none of these, SNAKE or HOSE BACK TOWARD HOUSE THROUGH INLET PIPE.

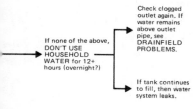

Check clogged outlet again. If water remains above outlet pipe, see DRAINFIELD PROBLEMS.

If none of the above, DON'T USE HOUSEHOLD WATER for 12+ hours (overnight?)

If tank continues to fill, then water system leaks.

If water has traveled enough distance to be clean, try curtain drain to divert clean water to an area where it won't cause erosion.

If water has traveled less than 100 feet, check soils. If coarse, then drainfield must be fixed. See DRAINFIELD PROBLEMS.

MAJOR DRAINFIELD PROBLEMS

Symptoms	Likely Causes	Solutions
Lush growth. Wet spots. Black liquid on surface.	Overloading (too many people, too small a system).	—Water conservation. —Additional drainfield —If black mat is present, new drainfield required. (Dig up drainfield area to 2' to 3' in a few spots. Look for black mat that stinks.)
Above plus steep slopes, clay soils, bedrock, etc.	Bad location and/or bad soils.	Alternatives to drain field (Chapters 3 and 8).
Works in dry season only.	High water table.	—Remove all runoff from roofs. —Conserve water. —Dig curtain drain and drainage ditches. —Consider waterless toilets and gray-water systems.
Nearby trees (and slow flush).	Clogging of drainfield pipes by roots.	Roto-Rooter pipes from septic tank.
Impermeable surfaces.	—Drainfield covered by patio or driveway.	Remove.
	—Water from surfaces flooding drainfield.	Dig curtain or peripheral drainage ditches.

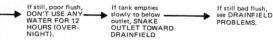

If still, poor flush, DON'T USE ANY WATER FOR 12 HOURS (OVERNIGHT).

If tank empties slowly to below outlet, SNAKE OUTLET TOWARD DRAINFIELD

If still bad flush, see DRAINFIELD PROBLEMS.

—Faulty installation is a strong possibility. Check the design. Is the septic tank in backwards? Are all the pipes connected? Are they collapsed? Is the septic tank or the drainfield too high above the house, sewer, or each other? Are all the necessary pipes perforated?

—The septic-tank system may have been designed for a smaller family or for weekend use. The size may be inadequate. Is this a cesspool (chapter 2) or truly a septic-tank system?

—The septic-tank system may have been constructed in a terrible location. Are you in a creek flood plain or a drainage swale? Are the soils any good?

Most problems are caused by a combination of conditions. In my town, a neighbor drove over the pipe between his house and the septic tank. The bent pipe worked fine until his son flushed his toothbrush down the drain. It caught in the bend. Another neighbor paved a patio over his drainfield. It worked fine all summer, but when the fall rains started the drains didn't drain nor the toilets flush. Many times water-table problems worsen minor mechanical failures.

When the pipe between the house and the septic tank is long and curved, grease accumulates at the turn, hardens, and can cause backing up into the home facilities. To declog pipes, first use Drano or Liquid Plumber. Don't use it again and again because these liquids can kill bacteria in your tank. Instead, use a plumber's snake or a hose with the nozzle turned on to the most forceful jet. (There are also commercial jet nozzles available.) The hose or snake should be twisted back toward the house from inside the septic tank. Another aid is called Drain King which seals pipe between hardened grease and nozzle, creating great pressure. Be careful with old pipes. Drain King is made by Scoville.

Detective work can be done with dyes. Your backyard may have puddles whose source is mysterious. Put the dye in the toilet or tank and see if the puddle is from the house, a broken water-main, a spring, or groundwater. Two common dyes are Uranine (Florascein) and Rhodamine B. Your sanitarian should have them. Rhodamine glows under a black light and is easy to see in small traces.

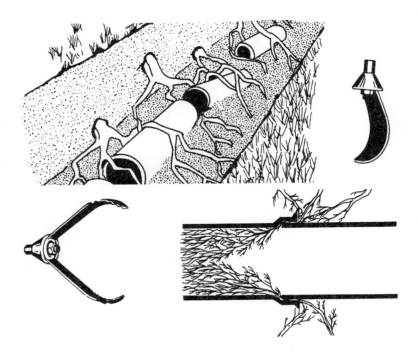

The Roto-Rooter has blades which cut roots or churn out soil that may be clogging your drainfield. They can be rented or you can hire someone to do it.

Reviving a Clogged Drainfield

Recent research has shown that it may be possible to revive a drainfield even after it has been clogged by the organic mat. To declog the soils in the drainfield, a 50 per cent solution of hydrogen peroxide is added directly into the drainfield. The hydrogen peroxide oxidizes the organic mat and reopens soil percolation. The peroxide treatment is a little complicated and dangerous because of the burning strength of peroxide fumes (goggles must be worn), but it is cheaper than a new drainfield, especially if there is no more room to excavate. Write to Industrial Chemicals Department, E. I. DuPont, Chestnut Run, Wilmington, Delaware 19898, or to POROX, c/o Marvin Woerple, Wisconsin Alumni Research Foundation, Madison, Wisconsin.

10. SEWERAGE AND COMMUNITY LIFE

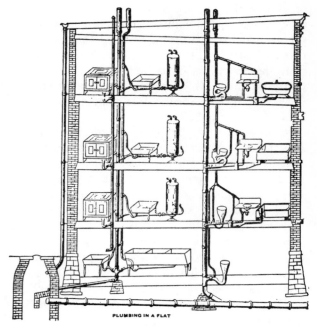

PLUMBING IN A FLAT

An apartment house in Chicago around 1901. Note how solids are held in a tank for removal and somewhat clarified effluent (wastewater) enters the sewer. This design still occurs in Japan. Pumped-out solids are sold to farmers for fertilizer.

In some towns, especially older ones, there are sometimes more septic-tank/drainfield problems than seem tolerable. The local health department does a survey and reports the percentage of failing tanks. Newspapers blare "HEALTH MENACE" or "SUR-FACING SEWAGE" and an editorial demands that the town sewer-up. Most often, it is totally unnecessary for the town to do so.

Most septic tank surveys confuse "failures" with problems of human neglect (like forgetting to pump). Rather than recommending the much cheaper alternative (repairing the problems) or investigating how many septic-tank systems need minor work or even major replacement, these surveys say: "Too many failures. Sewer up." Finally, the sanitarians who usually make the surveys have been working so hard that they have never read or taken classes in the new septic-tank/drainfield designs available or in the new tricks of the sanitary engineering trade (like evapotranspiration beds). These undereducated sanitarians can't even imagine home-site sewage treatment alternatives that would substitute for the Big Sewer.

Meanwhile, everybody gets railroaded by high-profit construction companies and super-tech engineering firms. Their representatives lobby the health departments, the utilities districts, and the government agencies concerned with water quality. They convince government agencies and townspeople that just another small grant would allow replacement of horrible septic-tank systems by a gold-Cadillac supersewer. There is actually a conflict of interest when an engineering firm (which makes profits on Big Sewers) is allowed to decide if septic tanks are working well. First, wearing the hat of consultant, the firm says the septic tanks are failing. Then, wearing the hat of engineer, the same firm says it will build the Big Sewer to replace the septic-tank systems. The consultant can hardly be considered unprejudiced if a fat contract will only occur when septic tanks fail. Laws should not allow the consultant firm to become the engineering/construction firm for the same project.

Many communities have been thoroughly boondoggled—especially by the technical lingo of "professional" consultants from the engineering firms, health fears, and the carrot of federal grants. Phrases like "mass failures," "inevitable progress," "economically feasible," or "ultimate dilution will lessen pollution" warp the straight-shooting, simple talk of early American town meetings. After the bond issue has passed, it is too late to stop the rapid housing development, tax increase, and, usually, added pollution caused by the Big Sewer. The town turns into suburban sprawl.

This chapter sketches the consequences on community life of the

Big Sewer and tries to arm health departments, government agencies, conservationists, and small communities with enough information to fight Big Sewers through the Environmental Impact Report now required for all large-scale sewer projects. It compares, point by point, the centralized sewer with at-home sewage treatment.

PHONY FAILURES

BUILD-OUT: A phrase frequently used by engineers and health departments to convince people that a centralized sewer is necessary. They assert that "science" says that "after 25 per cent of all the lots have been built upon, then too many failures occur and a sewer is needed." This kind of statement does not take into account the size of lots, their soils, or distribution of rainfall. It is not documented and is scientific nonsense.

INTERACTION FAILURES: The phony explanation used to explain how 25 per cent build-out causes septic-tank problems. Supposedly, there are so many drainfields, so close together, that they "interact" or interfere with each other. Supposedly, water cannot be drained away. These interaction failures are a total hype. Minimally, for drainfields to "interact," they must belong to next-door neighbors. Both neighboring drainfields should have problems. I have never seen one documented case of a community with many next-door neighbors with problem drainfields. Somebody created a dragon of fear to try to scare communities into an expensive sewer.

MASS FAILURES: Used to describe mass neglect. Only if soils are incredibly poor—like a subdivision built on a filled-in-swamp—might unrepairable mass failures occur. The failures would occur immediately or within two years. They would not be staggered—a few one year, a few the next. The only real failures are systems that can't be repaired or replaced.

There is a widespread belief in America that towns with home-site sewage treatment frequently have health disasters called "mass failures" or "interaction failures" because housing density is so thick that the soils can't absorb another drop of septic tank effluent. These stories are rarely true because one word—"failure"—has been twisted in a sneaky fashion.

Septic tanks can always be repaired or replaced or improved. A drainfield is really a failure only when it is unrepairable—when there is absolutely no room for a new or an artificially constructed drainfield. Any repairable drainfield is simply neglected and temporarily broken, not a failure. Almost all reported "failures" in health department or private surveys are simply a matter of neglect or ignorance. Even though the "failure" is repairable, these surveys call them failures as if they could never again be made to work.

So, Rule ⚹1: Always distinguish between a failure—a drainfield that cannot be repaired or replaced—and a problem which can be corrected.

Many bureaucracies, in pursuit of possible health hazards, try to insist that "failure" means (1) a septic-tank system with effluent daylighting above ground; (2) chronic backup of the toilet or drains; (3) slumping or seepage that causes erosion; or, less often, (4) records of pumping by septic-tank pumpers. But these "failures" are really symptoms of septic-tank problems and the need for repair. To jump from symptoms of a problem to abandoning septic tanks or drainfields altogether is like throwing out the baby because it has measles.

Rule ⚹2: Beware of being duped by stupid definitions. Symptoms indicate a need for concern, but not hopeless abandonment of your septic-tank system.

For instance, a town in California surveyed close to 400 septic tanks. The survey reported 12 per cent of the septic-tank systems showed symptoms of possible failure. But the town did not stop at the problem. It investigated the cause of each malfunction. It discovered that most of the causes were totally unrelated to clogged soils. (Clogged soils might require a new drainfield—often impossible on small lots.) Instead, the causes of malfunction were minor and easy to repair: broken pipes, neglected pumping, roots in the drainfield pipes, etc. By understanding the causes, two-thirds of the possible "failures" were working fine by the end of the survey. The newspapers didn't even carry the story.

Rule #3: Always distinguish the symptoms from the cause. The cause will determine whether a septic-tank system needs a minor repair or a major repair or is unrepairable.

In fact, almost all surveys report about 20 per cent "failures" (read: "problems"). This occurs regardless of soils, water use, design, or build-out. The "20 per cent" simply says that, throughout the United States, about 20 per cent of the residents have neglected to care for and maintain their home-site sewage treatment systems.

Finally, there is a world of difference between temporarily broken septic tanks and the ability of soils and drainage to allow septic-tank systems to function. There is no way to jump from the number of "failures" or "repairs" discovered by a survey to the ability of soils to provide adequate sewage treatment.

Rule #4: The ability of septic-tank/drainfield systems to work for a long life (fifty years) can only be known from an ecological survey of the soils, bedrock, water table, and evapotranspiration rates. A community with many "failures" may only be a community of maximum neglect.

Home-Site Sewage Treatment vs. The Big Sewer

The Big Sewer can be identified by a "collection system" which collects all the wastewater from many homes and funnels it into one big pipe ("the public sewer"). The public sewer then empties at a treatment plant, where some of the pollution and health hazard are reduced. Finally, the "treated" sewage is dumped out another pipe into a river, lake, or ocean.

Compared to the Big Sewer, home-site systems are very small scale. This smallness gives home-site systems their incredible superiority to the Big Sewer. Small is beautiful because it means more reliability, fewer pollution and sanitation hazards, lower costs, and lower exploitation of our energy resources (water and electricity).

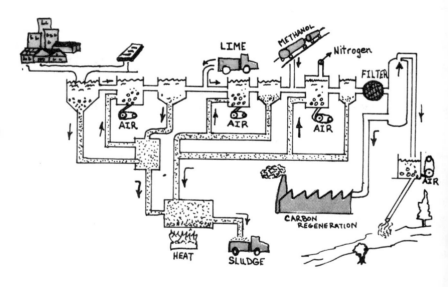

Modern high-tech ("tertiary") treatment plant showing extravagant use of energy (petroleum, methanol, lime, carbon, heat) to "dispose" of "wastes." There is still a sludge problem and costs and inefficiency are staggering. Methanol is used to get rid of nutrients without helping anybody or any agriculture.

Reliability

Maintenance and repair are the bogeymen of the Big Sewer. For instance, look at the typical "advanced" treatment plant shown on page 144 (tertiary treatment plant). The resulting water is pollution-free, but look at the number of mechanical parts needed to power the operation and the heavy dependence on transportation for the required chemicals and for sludge removal. To maintain such a system requires many skilled laborers and still presents a problem of what to do with the sludge.

Recently, sewage systems have also been plagued by strikes all along the line (truckers, chemical manufacturers, parts and equipment producers, plant operators). The combined cost of high wages and the need for petroleum power and electrical energy make maintenance and repair increasingly impossible. This is more true in small cities and large towns than in metropolises —money is tighter and experienced personnel are harder to find. In addition, as the *size* of the collection system grows, more sewers clog and overflow, more pipes break and need repair, and it becomes more difficult to keep drinking-water pipes separate from sewage pipes. (In New York City, seepages from one pipe to another are common.)

The unreliable performance of the Big Sewer has been tolerated for many years. American citizens have confused freedom from personal worry with reliability. "Out of sight, out of mind" has somehow come to mean, "Out of sight, no problems." Many Americans believed that giving responsibility for sewage treatment to a public authority automatically meant less pollution. But, as we all know, the transfer of responsibility to public authorities has *not* led to better treatment. Water pollution has become terrible. To make the situation worse, since a public authority does the work, citizens as well as sanitarians from county or state health departments may never hear of Big Sewer problems.

Americans are weird. They will accept sewage running down the street from a breached sewer as inevitable. But any suspicious odor or trickle from a neighbor's septic tank will send them running hysterically to the nearest health authority. In short, centralized sewerage problems have been covered up by public authorities and the American don't-want-to-think-about-it attitudes (the "excrement taboo"). Home-site sewage systems have been unjustly called "unreliable" by the same people who make money building the centralized sewerage systems, and until recently we all believed them.

Health and Sanitation

The fear of disease from sewage is totally exaggerated in the United States. The dragon of disease was alive when water supplies were drawn from rivers and lakes that were also used for sewage disposal. In the 1800s entire cities in Europe got typhoid or hepatitis. In 1977 there are very few water supplies that also allow sewage disposal. All water supplies are so highly chlorinated that no pathogens can survive. Almost every citizen is inoculated in childhood against the big killers: diphtheria and typhoid. Even sewage-plant workers contract water-borne diseases at a rate equal to everybody else.

The real water-borne diseases of the 1970s are practically ignored. They are caused by industrial pollutants like asbestos, petroleum byproducts, lead, and even fluorine and chlorine when combined with carbon-based molecules, all of which can cause cancer and blood disease. These can be prevented by drinking bottled spring water and stopping industries from polluting.

Yet, in spite of the accomplishments of modern medicine, the dragon of disease thrives in our imaginations. Although there has been no proven case of water-borne illness caused by septic tank effluent in more than ten years, health departments, engineers, and the newspapers still consider septic-tank systems more dangerous than the Big Sewer.

Home-site sewerage is safer than the Big Sewer—although, with heavy chlorination, both are pretty safe. Home-site sewerage is safer because of its small scale. The sewage quantity is small and treatment occurs in a small, confined area. The home site automatically *acts as a quarantine*—restricting the spread of disease if by chance the drainfield should have problems. The Big Sewer takes everybody's sewage and makes one big volume of it. The pathogens can multiply in this sewage, greatly increasing the chances of infection. The huge sewage systems are also more sub-

CODES AND HEALTH OFFICERS

County and state codes are more than mere minimum-design standards. Their original purpose was to prevent septic-tank systems from becoming a health hazard, to protect the consumer from contractors who might install badly designed and ultra-chintzy systems, and to reduce health-department workloads.

But the codes' original purposes have backfired. They have not been updated and they lack new understandings of home-site sewage treatment. State and county codes blandly copied federal guidelines which had mistakes. Many times, federal guidelines were misinterpreted—further codifying nonsense. Worst of all, health officers and sanitarians never had time to go back to school and receive further education about the new techniques and design features that could be used to improve home-site sewage treatment. Instead, the undereducated sanitarian had to interpret the codes rigidly rather than with an understanding of special ecological conditions. The result is that the codes themselves caused, in part, the failing septic-tank systems and the bad reputation of home-site sewerage.

The codes obviously need change. They need to address themselves to permanent home-site sewage treatment and to all kinds of home-site sewage systems. They need to be rewritten with built-in flexibility so that, for instance, water-conservation-minded homes could use compost privies. (At the moment, such privies are officially accepted only in Maine.) Health departments assume householders cannot learn to compost. Most important of all, county codes should have a built-in self-destruct provision so that they won't stay on the books long after they have served their usefulness.

Since the codes lost touch with ecological reality, they compensated by making design standards super fail-safe. This intentional overdesign makes for expensive septic-tank systems—so expensive that costs appear just slightly less than the Big Sewer. With government grants for centralized sewers, home-site systems soon disappear.

ject to breakdown, with consequences far worse than a problem drainfield's.

Home-site systems usually treat sewage in soils. The powers of soil to clean sewage water have been described in chapter 3. The Big Sewer tries to clean wastewater with many processes—only a very few use in-soil treatment. Usually, even after treatment, wastewater from the treatment plant must be heavily chlorinated. This kills pathogens and many other creatures. The killing of the "other" creatures has upset many natural balances in oceans and lakes. In other words, sanitation in home-style systems has no bad side effects, while the Big Sewer always kills more than just disease microbes.

Pollution and Resources

Again and again we have emphasized the recycling advantages of home-site systems—especially the recycling of nutrients in sewage to higher plants. Large-scale sewage treatment plants have incredible difficulty removing and disposing of nitrogen and phosphorus. To remove 90 per cent of the phosphorus (by flocculation with alum, lime, or ferric iron compounds) costs high-technology systems about $60 for every million gallons. The process, of course, kills protozoa and other creatures that are useful for self-purification.

Home-site systems remove phosphorus easily. One estimate states that about 210,000 tons of wastewater phosphates are recycled by home-site systems. This saves North Americans $100 million by avoiding the chemical processes of the Big Sewer, plus $800 million by avoiding the price of equivalent petroleum-based fertilizer.

In spite of this, there are still sanitary engineers and "respectable" scientists who argue—at the behest of private interest groups—that dilution of "wastes" is the solution to pollution. But, with rivers like the Hudson and lakes like Erie—and others all over the U.S.—becoming deader and deader, these scientists are sounding dumber and dumber. With the increasing costs of petroleum-based fertilizers, throwing "wastes" out in rivers sounds more and more ridiculous. Home-site systems are gems of man's ingenuity compared to Big Sewer inefficiency.

AERATOR

"Land retention" or recycling sewage treatment using the earth as a filter. It is a fine alternative to the big-techno-treatment plant shown on page 144. Note lack of mechanical parts (aerators are for emergency conditions). Each pond (in series) treats and re-treats wastewater. Side benefits are preservation of green space around cities, shorter distances for food transport, cheap agricultural water, and lower food prices. The best and easiest introduction is Clean Water, *by Leonard Stevens (1974, Dutton).*

Costs

Of course, most North Americans are most concerned about money, and it is in this arena that home-site systems offer the greatest advantage. At each stage of financing (initial costs, maintenance costs, replacement costs), home-site systems save money.

For instance, a survey of an area of 60,000 households in Kentucky estimated the *initial cost* of home-site systems to be $72 million. The cost of providing the same area with a Big Sewer and waste treatment was estimated at $354 million—a difference of $282 million! While the maximum cost of a home-site system was about $2,000 per home, the cost of the centralized sewer averaged between $2,500 and $6,000 per home. Some homes had to pay more than $18,000.

Maintenance costs are always very high for centralized sewage plants. You pay the public authority to pay wages to workers to watch over your sewage. You pay for all the electricity and chlorine that is minimally required to treat sewage. If treatment is advanced, you pay for special chemicals and filters. You pay for sludge disposal and, of course, all the upkeep for machines with many moving parts. A cared-for home-site system might have to be pumped once every four years at a cost of $30 to $85.

Replacement costs are equally ridiculous. If a community receives federal aid to build a treatment plant, it *must* pay the replacement cost—*while the treatment plant is wearing down.* At the end of twenty-five years, the community has twice paid for the treatment plant—once for the initial plant (in the form of taxes) and once for its replacement! This is called the Wastewater Capital Replacement Fund.

In addition, the Big Sewer lasts no longer than a well-designed and well-constructed home-site system. The federal government usually estimates the life of a Big Sewer as twenty-five years. A good septic-tank/drainfield system could last three times that long. Of course, there is a huge difference in scale for replacement costs. Big Sewer, big costs. Little at-home sewer, little costs.

American citizens have readily accepted big sewer costs because of our "Buy now, pay later" atmosphere. Like TV sets, homes, cars, or furniture, Americans are used to going into debt. The debt seems less bad when engineering firms and financial consultants say federal aid makes the Big Sewer "feasible." In addition, the debt is usually hidden inside taxes or other utility charges. Again, out-of-sight means no problem. This indebtedness has provided fast money for the sanitary engineers and left small towns struggling with their bills. Most debts, unless there is a new version of the Boston Tea Party, take forty years to pay.

The subsidizing of the Big Sewer is really a subsidy for the engineering and construction business. It increases their contracts. The government has never had a program to subsidize home-site systems (which it could) or encouraged home-site systems over centralized systems. There's no home-site lobby in Washington, D.C.

Community Effects of a Collection System

Centralized sewers have destroyed many American communities. Their immediate costs have forced retired citizens on fixed incomes and less wealthy citizens (especially the young) to migrate away from the increased sewer charges. The costs of bond issues, taxes, or sewer charges have polarized communities: pitting the haves against the have-lesses. Indirectly, the Big Sewer raises property taxes. The tax assessor assumes that a centralized sewer is an "improvement" over the home-site system. All "improvements" mean the property is worth more, and so your taxes go up.

Perhaps more important, collection systems encourage high-density development. Subdividers can squeeze more houses into the same piece of land when a collection system is installed. This changes a town's character and induces suburban sprawl. The Environmental Protection Agency, which gives grants for Big Sewers, discovered this in 1974. Its officers were shocked to find that its grants caused increased air pollution, subsidized private developers, promoted high population growth, and did not reduce water pollution. Their report is so important to future sewage and

land-use policies that many quotations are reprinted on the next page. The EPA is now (1977) restructuring its grants. One of its main concerns is to decrease water wastage that Big Sewers encourage. Obviously, the EPA will soon understand how home systems keep low-density villages, keep property taxes down, reduce water consumption, reduce pollution, and reduce costs. At any rate, all communities should be very aware that a centralized treatment plant and collection system can destroy the very life of a town.

Finally, the Big Sewer works against American freedom of choice. If a sewer runs by your house, you *must* hook up to it and pay the costs. In others words, you are not allowed to keep your home-site system, with all its advantages—even if it's working beautifully. This loss of option is killing the old American sense of self-reliance and responsibility. Undoubtedly, some backwoods Benjamin Franklin, unimpressed by the language of city-educated sewage experts, will soon stand up and say, "I won't." It will be a fine American court battle.

A New Alternative: On-Site Management Districts

In order to meet federal and state water-quality standards at the lowest possible cost and in order to use the best possible sewage treatment technology for each individual town, suburb, or small city, a new alternative is emerging in California and a few other states. A local agency (either the county or a special district) may own or simply regulate the home-site disposal systems of the community. This local agency (the "On-Site Waste Management District") keeps records of all septic-tank/drainfield locations, when they were pumped, what kinds of soil exist in the district, when and how new on-site systems should be built, what drainage problems affect drainfield function, etc. The on-site district will have the right to inspect septic tanks once every two or three years, require pumping if necessary, and require repairs to problems that are causing health hazards or pollution.

If public agencies begin to care for home-site systems, the bad reputation that somehow accompanies pit privies and septic tanks may change. Home-site sewage treatment can become permanent, sanitary, and a much-praised practice. The necessity and costs of high-tech systems should be reduced.

Interceptor Sewers and Suburban Sprawl:
|||||||||||||||||||||||||||||||||||||||||||||| Quotes from the E.P.A. ||||||||||||||||||||||||||||||||||||||||||||||

Current financing procedures—on both the local and federal level—may encourage the construction of sewerage systems tailored to the needs of future developers rather than the control of pollution problems. Where communities intend to finance the local share of project costs by connection fees on new development, this creates pressure to encourage rapid growth and thus ensure the financial viability of the new project . . .

Improve population forecasting techniques and review procedures. Population forecasts used in interceptors design should be better justified . . .

Use realistic standards for per capita flow. The frequent practice of encouraging engineers to size interceptors on the basis of a standard 100–125 gpcd water use measure should be replaced by employing actual water use statistics. There is evidence that the standard 100 gpcd

measure is excessive for many areas, and its use simply builds in additional excess capacity . . .

Encourage local communities to coordinate interceptor construction and land use planning and ensure that federal monies are used for their primary purpose—pollution control—and not the subsidy of low density residential development of vacant land . . .

Increase public participation in the planning process by publicizing community costs and benefits of interceptor-induced growth . . .

The design life of interceptors should be established at a maximum of twenty-five years, rather than at the current fifty year/ultimate population design period . . .

Do not provide federal funds for excess capacity. EPA should participate in financing only that portion of the interceptor project costs which represent the sewer capacity necessary to serve the needs of the existing population . . .

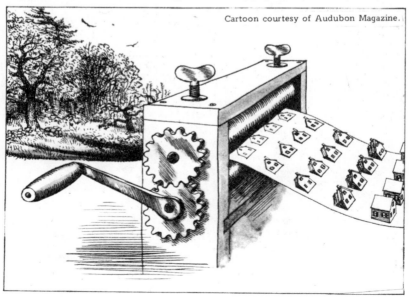

Cartoon courtesy of Audubon Magazine.

Fountain Run, Kentucky, considered various kinds of sewage treatment and disposal:

1. A centralized system with conventional sewers, ponds, and in-soil treatment and disposal.

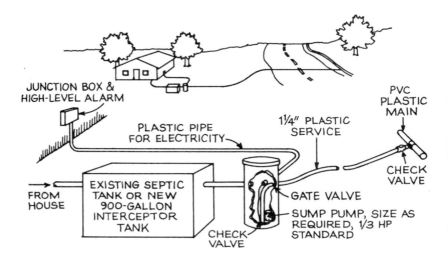

JUNCTION BOX & HIGH-LEVEL ALARM

PVC PLASTIC MAIN

PLASTIC PIPE FOR ELECTRICITY

1¼" PLASTIC SERVICE

FROM HOUSE

EXISTING SEPTIC TANK OR NEW 900-GALLON INTERCEPTOR TANK

CHECK VALVE

CHECK VALVE

GATE VALVE

SUMP PUMP, SIZE AS REQUIRED, 1/3 HP STANDARD

Pressure sewers are not new to the septic tank business. They are commonly used to pump septic tank effluent uphill to a better located drainfield. Now pressure sewers are becoming "standard" technology because they encourage water conservation (without large volumes of water some gravity sewers clog), because they can be cheaper than gravity sewers, and because they give flexibility to decentralized on-site sewerage (at least, to nearby off-site treatment). If a few existing houses have truly failing drainfields, their sewage can be pumped to a better location or septic tank truck—instead of sewering the whole town!

The pressure sewer is simply moving septic tank effluent with a pump through narrow diameter pipes to a trunk sewer, another drainfield or a holding tank for truck disposal.

2. A decentralized system with some septic tanks feeding community drainfields (e.g., five houses per drainfield) and some individual drainfields (on good soils).
3. A totally *home-site system* with every septic tank
having an at-home individual drainfield (on poor
soils, the drainfield would be specially engineered).

The federal government requires a comparison to be
made of all alternative sewage systems. The comparison
showed that the centralized system would cost the home-
owner 2.4 times as much as the decentralized system and
three times as much as the home-site system. The centralized system also was 25 per cent more damaging to the
environment (in terms of soil loss, energy consumption,
stream damage, groundwater table, pollution from development, etc.) than either the decentralized or the home-site systems.

At a public meeting, Fountain Run residents discussed
the alternatives. Centralized sewage (the "Big Sewer") was
so expensive ($17 per month, with a 75 per cent grant, or
$38 per month without a grant) that only a few homes could
afford it. Home-site systems, though cheapest ($6 with
grants; $13 without grants), forced homes on poor soils to
pay high prices for specially engineered systems. The decentralized plan, with the combination of home-site systems and
communal drainfields where necessary, was favored ($7 per
month with grants, $17 without grants).

The decentralized sewage plan was so strange to state
officials that they ranked it 240 out of 241 projects on their
priority list. (Septic tank malfunction was not included in
their ranking scheme!) It seemed that the state would rather
wait until there was no room for decentralized sewage treatment. Then, as a top priority, the state would support an
overly expensive Big Sewer with worse treatment and higher
costs. Federal officials from the Environmental Protection
Agency said the project would not be reviewed for ten years!
This is poor preventive medicine.

Two problems remain: Will the federal government make grants available to public agencies to upgrade, maintain, and construct home systems? They give millions to centralized sewage, but for some unknown reason balk at giving grants for home-site systems. Secondly, who has the authority to regulate home sewage systems? Is it the local, the county and/or state health officials, or the utility or sanitary districts? What are the legal limits on public authorities' checking septic tanks on private property (like the gas man or the electric man inspecting the meter)?

In the next five years, we will perhaps see these questions brought to court, legislatures, and Congress. The results will establish the future precedents for home-site over centralized sewerage. (See: "Rural Wastewater Alternatives" in the Bibliography.)

Summary

Home-site sewage treatment is cheaper, pollutes less, recycles more, slows or controls suburban sprawl, has fewer health hazards, and remains personal and intimate with the necessities of water, nutrients, and the lives of other creatures. Centralized sewage disposal, shielded by public authorities, has kept citizens unaware of sewage costs and inadequate treatment and disposal, as well as of their own natural responsibility for recycling their own wastes and keeping other plants and animals productive and healthy.

SEEPAGE PIT CONSTRUCTION

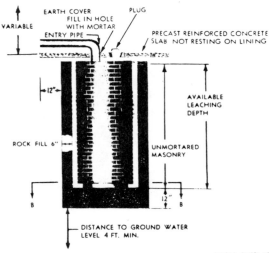

VARIABLE

EARTH COVER
FILL IN HOLE
WITH MORTAR

PLUG

ENTRY PIPE

PRECAST REINFORCED CONCRETE
SLAB NOT RESTING ON LINING

12"

AVAILABLE
LEACHING
DEPTH

ROCK FILL 6"

UNMORTARED
MASONRY

B

12"

B

DISTANCE TO GROUND WATER
LEVEL 4 FT. MIN.

BRICKS OVERLAP ON EACH LAYER

SECOND LAYER

THIRD LAYER
FIRST LAYER

SECTION A-A

PLACE 6" COARSE AGGREGATE
(½" TO 1") AROUND UNMORTARED
MASONRY

BRICKS LAID
CLOSE WITHOUT
MORTAR

FIRST LAYER OF BRICK
SECTION B-B

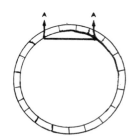

A   A

SECOND LAYER OF BRICK

NOTE: SECOND AND REMAINING
LAYERS ARE LAID END TO END
AND AT RIGHT ANGLES WITH FIRST
LAYER OF BRICK

SUGGESTED SPECIFICATIONS FOR WATERTIGHT CONCRETE

1. *Materials*

Portland cement should be free of hard lumps caused by moisture during storage. Lumps from dry packing that are easily broken in the hand are not objectionable.

Aggregates, such as sand and gravel, should be obtained from sources known to make good concrete. They should be clean and hard. Particle size should range from very fine to ¼ inch.

Gravel or crushed stone should have particles from ¼ inch to a maximum of 1½ inches in size. Water for mixing should be clean.

2. *Proportioning*

Not more than 6 gallons of total water should be used for each bag of cement. Since sand usually holds a considerable amount of water, not more than 5 gallons of water per bag of cement should be added at the mixer when sand is of average dampness. More mixing water weakens the concrete and makes it less watertight. For average aggregates, the mix proportions shown in the table at right will give watertight concrete.

3. *Mixing and Placing*

All materials should be mixed long enough so that the concrete has a uniform color. As concrete is deposited in the forms, it should be tamped and spaded to obtain a dense wall. The entire tank should be cast in one continuous operation if possible, to prevent construction joints.

4. *Curing*

After it has set, new concrete should be kept moist for at least seven days to gain strength.

Type V portland cement may be used when high sulfate resistance is required.

	Average Proportions for Watertight Concrete			
Max. Size Gravel (in.)	Cement (vol.)	Sand (vol.)	Gravel (vol.)	Water[1] (vol.)
1½	1	2¼	3	¾
¾	1	2½	2½	¾

[1] Assuming sand is of average dampness.

MANUFACTURERS AND DISTRIBUTORS OF
WATER-CONSERVATION PRODUCTS

* ALSONS PRODUCTS
CORPORATION
525 E. Edna Place
Covina, California 91722
    Hand-held shower

* AMERICAN STANDARD-
PLUMBING/HEATING
P.O. Box 2003
New Brunswick, New Jersey 08903
    Shallow-trap toilets; self-clos-
    ing valves; pressure-balancing
    mixing valve; shower fixture,
    Aquarian (4 gpm) flow con-
    trol, Aquamizer (2.5 gpm)

* AQM CORPORATION
1909 New Rodgers Road
Levittown, Pennsylvania 19056
    "Aqua-miser" toilet-tank insert

‡ AQUA-DATA CORPORATION
P.O. Box 901
Carpinteria, California 93013
    Drip irrigation

* AQUA GUARD
3200 Valley Lane
Falls Church, Virginia 22044
    Toilet-tank insert

† AQUA MISER
P.O. Box 284
Glen Ellen, California 95442
    Aqua Miser toilet-tank insert

* BRIGGS
5200 West Kennedy Boulevard
P.O. Box 22622
Tampa, Florida 33601
    Conserver, shallow-trap toilet;
    conventional toilets; bidet

‡ CHAPIN WATERMATICS, INC.
368 N. Colorado Avenue
Watertown, New York 13601
    "Micro-Dripper," drip irriga-
    tion emitter, direct trap into
    polyethylene pipe

* CHICAGO FAUCETS
2100 S. Nuclear Drive
Des Plaines, Illinois 60018
    Econo-flow (0.75 gpm aera-
    tor) faucet; Soft-flo flow con-
    trol; Stedi-flo flow control;
    self-closing faucets; conven-
    tional faucets

* COLT INDUSTRIES
Water and Waste Management
Operation
701 Lawton Avenue
Beloit, Wisconsin 53511
    Envirovac system vacuum toi-
    let; Liljendahl, Electrolux

* CONSERVOCON, INC.
191 Edgewater Street
Staten Island, New York 10305
    Foot controlled faucet ($9
    attachment)

* Flushing devices, faucets, and showers
† Miscellaneous water-saving devices
‡ Irrigation devices

NOTE: The inclusion of a company's name in this list does not in any
way constitute an endorsement of its products.

‡ CONTROLLED WATER
EMISSION SYSTEMS
585 Vernon Way
El Cajon, California 92022
    Drip irrigation

* CRANE COMPANY
17900 Skypark Circle
Irvine, California 92664
    Shallow-trap toilet

‡ DUPONT COMPANY
1007 Market Street
Wilmington, Delaware 19898
    Drip irrigation

‡ DEEP SEEP CAP-TOP
SOAKERS
915 E. Bethany Home Road
Phoenix, Arizona 85014
    Drip irrigation

‡ DEFCO, INC.
325 N. Daloson Drive
Camarillo, California 93010
    Emitters for drip irrigation

† L.M. DEARING ASSOCIATES,
INC.
12324 Ventura Boulevard
Studio City, California 91604
    Floating pool cover

‡ DIXEL IRRIGATION SYSTEMS
17 Briar Hollow
Houston, Texas 77027
    Drip irrigation—strap-on emit-
    ter

* DUO-FLUSH PLUMBING
COMPANY
610 S. Tejon
Colorado Springs, Colorado 80903
    Two-way flushing valve

* EATON CORPORATION
Controls Division
Plumbing & Heating Products
191 E. North Avenue
Carol Stream, Illinois 60187
    Dole flow controls (2, 3, 4,
    gpm); Dole shower controls
    (2, 2.5, 3, 4 gpm); Dole sink
    faucets (1.5 to 4 gpm models)

* ECOLOGICAL WATER
PRODUCTS INC.
(EWP)
P.O. Box 509
Dunkirk, New York 14048
    Nova shower head (2.5 gpm);
    EWP faucets aerators (1.5
    gpm)

* ECOLOGY PLUS
Box 184
Crydon, Pennsylvania 19020
    Water Wizard tank insert

* ELJER PLUMBINGWARE
Wallace Murray Corporation
Three Gateway Center
Pittsburgh, Pennsylvania 15222
    Savon water-saving urinal;
    shallow-trap toilet conven-
    tional urinal; conventional
    toilet

* RICHARD FIFE, INC.
140 Greenwood Avenue
Midland Park, New Jersey 07432
    Spray taps; flow control built
    into the faucet; Rada thermo-
    static mixing valve

* FLUIDMASTER, INC.
P.O. Box 4264
Anaheim, California 92803
    Tank-flushing valve, leak sig-
    naling ballcock

* FORMULABS, INC.
Fluorescent Dye Tracing Systems
Division
529 W. Fourth Avenue
P.O. Box 1056
Escondido, California 92025
    Water-saver super drip kit;
    toilet-tank dye tablet

* GENERAL AMERICAN
TRANSPORTATION
CORPORATION
General America Research
Division
7449 N. Natchez Avenue
Niles, Illinois 60648
    Controlled-volume flush toilet

‡ HYDRO-RAIN
26031 Avenida Aeropuerto
San Juan Capistrano, California
92675
    Solenoid valves for lawn
    sprinkling and irrigation

‡ HYDRO TERRA
CORPORATION
800 N. Park Avenue
Pomona, California 91768
    Automatic irrigation control-
    ler tensiometers

* INTERBATH, INC.
3231 N. Durfee
Elmonte, California 91732
    Hand-held shower

* ITT LAWLER
453 N. MacQuesten Parkway
Mount Vernon, New York 10552
    Pressure-reducing valve house
    connection

† IN-SINK-ERATOR (I.S.E.)
4700 21st Street
Racine, Wisconsin 53406
    Instant hot-water heater

† INTERNATIONAL METAL
PRODUCTS
Division of McGraw Edison
Box 20188
500 S. 15th Street
Phoenix, Arizona 85036
    Evaporative air cooler

‡ IRRI-DRIP SYSTEMS,
INC.
P.O. Box AW
Ventura, California 93001
    Drip irrigation

* JEF SKID
P.O. Box 2288
Rockville, Maryland 20852
    Dual-flush toilet

* JKW 5000 LIMITED
10610 Culver Boulevard
Culver City, California 90231
    Water-gate toilet-tank insert;
    Ny-Del shower head.

† KITCHEN AID
Hobart Manufacturing Company
1501 W. 8th Street
Los Angeles, California 90017
    Instant hot-water heater

* KOHLER COMPANY
Kohler, Wisconsin 53044
    Water Guard shallow-trap toi-
    let; pressure-balancing mixing
    valves; flow-control shower
    heads (3 gpm and up); flow-
    control faucets (2 gpm and
    up)

‡ LEISURE TIME WATERING
SYSTEMS
P.O. Box 1298
Hollister, California 95023
    Drip irrigation

† LIFTOMATIC, INC.
6445 N. Sepulveda
Los Angeles, California 90049
    Swimming-pool covers

* MANSFIELD SANITARY, INC.
Perrysville, Ohio 44864
  Water-saver toilet flush valve;
  vacuum flush toilets for boats

* METROPOLITAN
WATERSAVING COMPANY,
INC.
5130 MacArthur Boulevard, Suite 106
Washington, D.C. 20016
  "Little John" water-closet insert

* MICROPHOR
P.O. Box 490
Willits, California 95490
  Microphor pressurized flush toilet (2 quarts per flush)

* MINI-DAM WATERSAVER
CORPORATION
640 S. Pickett Street
Alexandria, Virginia 22304
  Toilet-tank insert (see Aqua Guard)

* MINUSE ENVIRO-SYSTEMS,
INC.
206 N. Main Street, Suite 300
Jackson, California 95642
  Shower (2 quarts per minute)

* MOEN
Elyria, Ohio 44035
  Pressure-balancing valves;
  Easy-clean shower head (3 gpm); aerators

* NOLAND COMPANY
2700 Warwick Boulevard
Newport News, Virginia 23607
  Flow controls (plastic inserts)

* NY-DEL CORPORATION
740 E. Alosta Avenue
Glendora, California 91740
  The "Saveit" water-closet insert

† OWENS/CORNING
5933 Telegraph Road
Los Angeles, California 90040
  Hot-water-pipe insulation

‡ PACEMAKER CORPORATION
3828 Fifth Avenue
San Diego, California 92103
  Moisture indicator

† PIPER HYDRO, INC.
1159 Fountain Way
Anaheim, California 92806
  Combined domestic hot-water-circulation and heating system

* POWERS REGULATORY
COMPANY
Skokie, Illinois 60076
  Pressure-balancing shower and bathtub valves

* PROGRESSIVE HARDWARD
P.O. Box 874
Garden Grove, California 92642
  Ballcock

‡ RAIN BIRD
Glendora, California 91740
  Drip irrigation emitters

* ROKAL ARMATUREN GMBH
4053 Nettetal–1
Postfach 1266
West Germany
  Thermostatically controlled mixing valve

* SAVE IT OF WASHINGTON
11168 Safford Way
Reston, Virginia 22070
  Toilet-tank inserts (2 models)

‡ SOILMOISTURE EQUIPMENT
CORPORATION
Box 30025
Santa Barbara, California 93105
  Soil tensiometers

* SPEAKMAN COMPANY
Wilmington, Delaware 19899
    Flow controls, Auto flow (2.5
    to 4.5 gpm); spring and time
    faucets; shower heads, Auto
    flow (2.5 to 4.5 gpm)

‡ SPOT SYSTEMS, INC.
P.O. Box 807
Redmond, Washington 98052
    Drip irrigation emitters

‡ SUBMATIC IRRIGATION
P.O. Box 2449
Menlo Park, California 94025
    Drip irrigation

† SUNSET POOLS
900 Wilshire Boulevard, Suite 1242
Los Angeles, California 90017
    Solar circles (floating swim-
    ming-pool "solar heaters" and
    evaporation reducers)

* THETFORD CORPORATION
Waste Treatment Equipment
Division
P.O. Box 1285
Ann Arbor, Michigan 48106
    "Cycle-let"   recycling   flush-
    water toilet

‡ TURF SERVICE LABORATORY
P.O. Box 1001
Laguna Beach, California 92651
    Moisture indicators, moisturom-
    eter

‡ TYME VALUE CORPORATION
12100 E. Park Street
Cerritos, California 90701
    Sprinkler control

* UTAH MARINE
459 S. Seventh Street
P.O. Box 485
Brigham City, Utah 84302
    Flush   Gard,   sink-bob   dual
    flush device

This illustration shows the
upholstered commode open.

‡ WAICO-NORTHWEST
5920 N.W. 87th Avenue
Portland, Oregon 97220
    Drip irrigation

* WATER CONTROL
PRODUCTS, INC.
1100 Owendale, Suite E
Troy, Michigan 48084
    Flushmate pressurized tank
    toilet

* WATTS REGULATOR
COMPANY
Lawrence, Massachusetts 08142
    Water-pressure-reducing valve
    house connection

* WRIGHTWAY
MANUFACTURING
COMPANY
371 E. 116th Street
Chicago, Illinois 60628
    Water-saver kit (shower head
    and two faucet aerators)

BIBLIOGRAPHY |||||||||||||||||||||||||||||||||||||||||||||||||||||||||||||||||||||||||||||||||||||||||||||||||||||||||||||||||||||||

Septic Tank Practices

Current and Recommended Practices for Subsurface Waste Water Disposal Systems in Arizona, by J. T. Winneberger et al. 1973. $3.50. From the School of Engineering Sciences, Arizona State University, Tempe, AZ 85281.
This is the definitive encyclopedia of septic-tank practices. Good advice on everything: percolation tests, use in the United States, bad surveys, good designs, proper management, government headaches, privies, and sewers. A must for every health department, engineering firm, and home-site sewerage connoisseur.

Treatment and Disposal of Waste Water from Homes by Soil Infiltration and Evapo-Transpiration, by Alfred P. Bernhart. 1973. 173 pp. in vol. I. University of Toronto.
No cost analysis. No maintenance discussion. A little too high-tech for my taste. Nevertheless, the best overall book on all other aspects of anaerobic and aerobic drainfield and septic-tank design. Especially good on how living creatures clean wastewater.

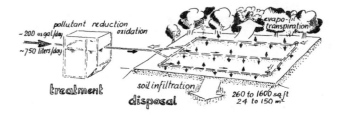

Alternate Sewage Manual, From P.O. Box 309 Madison, WI 53701. Price unavailable.
The best written descriptions of mound drainfields using septic tank and dosing chambers. Three parts: impermeable soils, shallow soils with porous bedrock, and high groundwater. Excellent discussions of pumping size, etc. A little high-tech.

On-Site Sewage Disposal, by Jack L. Abney. Private publication, $1.00 welcomed. From Air Pollution Control Department, Room 207, Civic Center Complex, Evansville, IN 47708.
The best short design manual for on-site systems, especially "mound" systems.

Water

Residential Water Conservation, by Murray Milne. 1975. 468 pp. $7.50 postpaid from California Water Resources Center, University of California, Davis, CA 95616.

This book replaces all other books on the subject. An excellent history of attitudes towards water conservation and of changing technology; a cross-section of typical American water use; dozens of devices for home water conservation including a survey of all alternative toilets; how water districts can encourage or discourage water wastage; a huge annotated bibliography and directory to manufacturers.

Demonstration of Waste Flow Reductions from Households, by S. Cohen and H. Wallman. No. PB 236 904/AS. $5.25. From National Technical Information Service, U. S. Dept. of Commerce, Springfield, VA 22151.

Herein lies the vision of a saner America. Good technical know-how on dual-flush toilets, greywater recycling, shallow-trap toilets, flow-reduction valves, and more. Buy a copy; copy pertinent parts for friends.

Conference on Water Conservation and Sewage Flow Reduction with Water-Saving Devices, Proceedings. William E. and Peter Sharpe, editors. April 1975. From Pennsylvania State University, University Park, PA 16802.

Technical papers on all aspects: economics, field experiences, education, hydraulics, current trends in the United States.

The Algal Bowl, by John R. Vallentyne. $4.32. From Information Canada, Ottawa, Canada K1A 0S9.

A great book about lakes and eutrophication—the overstimulation of lakes with detergents and other human wastes.

Consciousness

Farmers of Forty Centuries, by F. H. King. 1911. 441 pp. $9.95 post-paid from Rodale Press, Inc., 33 East Minor, Emmaus, PA 18049, or Whole Earth Truck Store, 558 Santa Cruz, Menlo Park, CA 94025.
A beautifully written travelogue on the agriculture and the recycling of all wastes that have been perfected through the forty centuries of Chinese, Korean, and Japanese attentiveness.

Clean Water, by Leonard Stevens. 1974. 289 pp. From E. P. Dutton & Co., N.Y., or Whole Earth Truck Store. Hardback, $10.00. Will be available in paper.
The first history of sewage farms and the great ability of the earth to clean water. Lots of fine examples of existing recycling systems. Gives hope in an America of extravagant waste. Beautifully written.

Biological Control of Water Pollution by Joachim Tourbier and Robert Pierson. 1976. $20. From: Center for Ecological Research, Dept. of Landscape Architecture and Regional Planning, University of Pennsylvania, 3933 Walnut St., Philadelphia, PA 19174 or Whole Earth Truck Store.
The use of living tissue as filter, chemical renovator and strainer to clean and recycle polluted water. The best overview. Too expensive.

On-Site Waste Management, vol. II, by J. T. Winneberger. 52 pp. Free. From Hancor, Inc., Findlay, OH 45840.
Two essays: "Pertinent Points in the Nitrogen Cycle" and "Setback Needed to Protect Water Supplies from Viruses." Both reduce fear levels to reasonableness. The nitrogen essay indicates actual quantities and how to determine them for waste management. The viral essay turns worries into soil-adsorption analysis. Great bibliography.

"The Fantasy of Dirt," by Lawrence Kubie, in *The Psychoanalytic Quarterly,* Oct. 1937.
The fantasy that dirt is anything which emerges from the body or has contact with a body opening. As if the body itself creates filth at every aperture. Kubie tries to understand American Mind—to free it.

Community

Land Use and the Pipe, by R. D. Tabors, et al. 1976. 180 pp. D. C. Heath and Co., Lexington, MA.
Study showing that government financing is encouraging growth in communities that did not want it, and that citizens must pay for this future growth by paying for superlarge systems and systems designed for extravagant water use. Perfect document to fight a sewer.

Wastewater Treatment for Small Communities, by George Tchobanoglous. 1973. 44 pp. University of California, Davis, CA 95616.
The best summary for small communities with between 5,000 and one million gallons per day of dry-weather sewage flow. Includes Imhoff tank, tricking filter, activated sludge, ponds, and land disposal. Good on operations and maintenance—a big problem for financially tight towns.

On-Site Waste Management. Vol. VI. Free. From Hancor Inc., Findlay, OH 45840.
Two essays: "On-Site Wastewater Management Districts," by J. T. Winneberger and James Burgel, and "Pressure Sewers" by J. Burgel. Winneberger wrote the pivotal essay on On-Site Management Districts ten years ago. Much of the essay is in the OAT volume but this is the real McCoy. Similarly, the pressure sewer essay is the best summary available although parts can be found in the OAT volume.

"Sanitary Surveys and Survival Curves of Septic Tank Systems," by J. T. Winneberger, in *Journal of Environmental Health,* vol. 38, no. 1 (July–Aug. 1975), pp. 36–39.

Integration of On-Site Disposal in a 201 Facilities Plan, by Jack Abney. 1976. From Parrot, Ely and Hurt Consulting Engineers, Inc., 620 Euclid Ave., Lexington, KY 40502.
The story, in detail, of Fountain Run, described briefly on page 154.

Other Home-Site Sewerage

Rural Wastewater Disposal Alternatives, 1977; 145 pp. Free from: State Water Resources Control Board, P.O. Box 100, Sacramento, CA 95801.
This report by California's Office of Appropriate Technology (OAT) connects governmental regulations to home-site systems and greywater. It is a "state-of-the-art" report with descriptions of all the home-site systems, health problems, needs for research, and possibilities of On-Site Waste Management Districts. A must for every County and State Health Department.

Goodbye to the Flush Toilet, Edited by Carol Hupping Stoner, 1977. 285 pp. $6.95. From Rodale Press, Emmaus, PA or Whole Earth Truck Store, 558 Santa Cruz, Menlo Park, CA 94025.
More popularly written than the OAT report and with more kinds and much better diagrams of waterless toilets and greywater systems. Includes a fine essay on how the flush-and-forget mentality as well as how water-borne pollution came to be.

Excreta Disposal for Rural Areas and Small Communities, by E. G. Wagner and J. N. Lanoix, 1958. 187 pp. $8.50 plus 75¢ handling. Available from Q Corporation, 49 Sheridan Ave., Albany, NY 12210, or Whole Earth Truck Store, 558 Santa Cruz, Menlo Park, CA 94025. Still the best book on waterless toilets of all kinds and how to construct them. The Farallones Compost Privy (page 16) is a modification from this book. Published by World Health Organization with a fine pan-cultural attitude.

Low Cost Technology Options for Sanitation, by Witold Rybczynski, Chongrak Polprsert, and Michael McGarry, 1978, IRDC Pub. No. 102. From: UNIPUB, Box 433, Murray Hill Station, N.Y., N.Y. 10016.
A bibliography for third world nations. Indispensable for those interested in working with the new technology of on-site collection and treatment (latrines, privies, septic tanks); off-site collection (cartage and waterborne) as well as low-tech treatment and re-use (ponds, irrigation, aqua-culture, algae harvesting, biogas, composting). Short biblio on greywater and water conservation.

Soils and Plants

Soil Survey Manual. USDA Handbook No. 18. 1951. U. S. Government Printing Office, Washington, DC 20401.
Soils are the crucial focus: You must squeeze them, roll them, dig them, contemplate their colors. This book is far and away the best for on-site sewerage students and farmers.

Nature and Properties of Soils, by H. Buckman and N. Brady (7th ed.). 1969. 650 pp. MacMillan, N.Y.; available from Whole Earth Truck Store, 558 Santa Cruz, Menlo Park, CA 94025.
The technical text on soils with good solid facts.

Biology of Plants, by Peter Raven and Helena Curtis, 1970. 706 pp. Worth, N.Y.; available from Whole Earth Truck Store.
A college text of great beauty on every aspect of plant life: DNA to the tropical jungle.

KEEPING UP-TO-DATE

Rain Magazine. 2270 NW Irving, Portland, OR 97210. $10/year.
The best reviewing journal of new developments in all aspects of appropriate, small-scale technology and innovative thoughts for the future.

Co-Evolution Quarterly. Box 428, Sausalito, CA 94965. $12/year.
A wide-ranging journal with essays, photos and interviews covering more areas of American Life than any other publication going. The author of this book reviews sewage and water and soil topics for *CQ*.

Technical

A Study of Methods of Preventing Failures of Septic Tank Percolation Systems, by J. T. Winneberger and P. H. McGauhey. 1965, SERL Report No. 65-17. Sanitary Engineering Research Lab, University of California, Berkeley, CA 94720.

Soil Mantle as a Waste Water Treatment System, by P. H. McGauhey and R. B. Krone. 1967. SERL Report No. 67-11. Sanitary Engineering Research Lab, University of California, Berkeley, CA 94720.

"The Concept of an Infiltration Quotient Opens Up Possibilities for Subsurface Field Design," by J. T. Winneberger, in *Journal of Environmental Health,* vol. 38, no. 2 (Sept.–Oct. 1975), pp. 113–18.

"The Correlation of Three Techniques for Determining Soil Permeability," by J. T. Winneberger, in *Journal of Environmental Health,* vol. 37, no. 2 (Sept.–Oct. 1974), pp. 108–18.

Wastewater Engineering: Collection, Treatment, Disposal, by Metcalf and Eddy, Inc. 1972. McGraw-Hill Book Co., N.Y.

Sanitary Significance of Fecal Coliforms in the Environment, by E. E. Geldreich. Federal Water Pollution Control Administration, Pub. WP-20-3, Nov. 1966.
Explains the many sources of coliforms from humans, ducks, and even fish and the confusions of most health departments. A book that lowers fear levels.

ACKNOWLEDGMENTS

Special (third edition) love for Joanne. Tim Winneberger and Greg Hewlett are practically the authors—especially by influence. This book has a beautifully entangled history from next-door neighbors to federal officials. My neighbors deserve most credit for talking greywater, suggesting alternatives, and helping me learn by helping them. Lloyd Kahn provided the spirit, direction, and example for all this to come about. Minor Wilson provided many drawings, the space, and music to keep it all going. Joe Bacon composed the first two editions. Many, many friends helped with collating, stapling, and binding the early home-grown versions. Sarah Hammond and Jane Bacher taught me invoicing. Micky Cummings, the printer of Mesa Press, turned negatives to plates and plates to pages. Very smooth. Finally, Arthur Okamura did three delightful covers—for the embryonic edition, the neighborhood survey, and the Mesa Press edition.

In its early stages, important financial assistance came from Point Foundation. The *Co-Evolution Quarterly* provided their office composer twice. Thank you, Stewart Brand. Stewart also turned John Brockman on to this booklet, and he gave it to Bill Strachan of Doubleday. I thank them for helping me carry this information and conservation consciousness to many more people than I would have ever dreamed possible.

CREDITS

Algal Bowl (see Bibliography): 6, 7.

Japanese Homes and Their Surroundings by Edward Morse: 13.

Excreta Disposal for Rural Areas and Communities (see Bibliography): 14, 19, 61 (sandbox).

Goodbye to the Flush Toilet (see Bibliography): 18 (compost pile), 56 (pressure toilet), 63 (Clivus filter), 70, 162.

Biology of Plants (see Bibliography): 30 (stomata), 35.

Marin County Water Conservation Guide: 46, 50, 51.

Residential Water Conservation (see Bibliography): 52, 53, 55, 57, 58.

Segregation and Separate Treatment of Black and Grey Household Wastewaters to Facilitate On Site Surface Disposal by Robert Siegrist, Small Scale Waste Management Project, University of Wisconsin: 63 (sand filter), 98.

Don Ryan: 59, 60, 61.

Sensitive Chaos by Theodore Schwenk: 67.

Home Sewage Disposal (Special Circular 212) by Penn State University, College of Agricultural Extension Service, University Park, PENN: 86, 92 (top), 130, 133, 137.

Geology for Individual Sewage Disposal Systems by Alvin Franks (California Geology, Sept. 1972): 89.

Maine Sewage Disposal Manual: 86 (top), 96 (top).

Treatment and Disposal of Waste Water from Homes (see Bibliography): 114, 165.

Building a Home in the Country? by R. J. Otis (Small Scale Waste Management Project, ibid.): 116.

On-Lot Subsurface Sewage Disposal Systems by Edward Palmer, College of Agriculture, University of Connecticut, Storrs, Conn: 124.

Dan O'Neill: 151.

Nostalgia: 157.

Audubon Society: 153.

ABOUT THE AUTHORS

During the 1971 San Francisco oil spill, Peter Warshall joined many others in an attempt to clean about 3,000 oiled, oceanic birds. All died. During the spill he met Greg Hewlett. They (with others) discussed saving community life—human and all other creatures. A town nearby was polluting the ocean with sewage and with their new interest, they began to probe. The local Utilities District had hired engineers to solve the sewage problem. Their answer was incredible: Combine the town's sewage with a neighboring town's, add all the people who still had septic-tank systems to the Big Sewer, and then dump it all further out in the ocean.

Since the cure seemed worse than the disease, Greg, Peter, and others followed their vision to save the Pacific. They called the Environmental Protection Agency, the County Health Department, the State Water Quality Control Board. They learned words "outfall," "effluent," "interceptor sewer," "primary," and "secondary treatment." They wrote letters, attended Utilities District meetings, finally challenged the "expert" engineers and biologists. Most of all they tried to convince everybody they met that the solution to pollution was *not* dilution in the ocean.

Simultaneously, other townspeople began to notice the consequences of the Big Sewer: increased taxes and sewer charges, increased human population and decreased animal and plant life, increased pollution from the increase in people, forced exodus of townspeople with fixed or low incomes, etc. A combination of financial and ecological worries inspired a little-used American option: the recall election. (If 10% of the registered voters want, an elected official can be forced back into running again or quitting.) The concerned towns ran recall elections. The anti-Big Sewer crowd won. Greg and later Peter were elected to the Utilities District.

The State Water Quality Board responded by decertifying the Big Sewer—even though they had previously given the Big Sewer their blessings. The EPA said a new Impact Report would be required which would take six months. The Big Sewer "expert" engineers and biologists smelled a sinking gold Cadillac and backed off.

The newly elected officials of the Utilities District still had a pollution problem. Enter the third author—Dr. T. J. Winneberger. He explained about septic tanks and drainfields, about lousy vs. intelligent surveys. He convinced the part of the town with septic tanks to stay with these beautiful, small-scale recyclers. Meanwhile, the sewered part of town that was dumping sewage directly into the ocean designed a recycling system like the one shown on page 149. Land disposal meant all nutrients would be used to grow plants and the land required would be safe from subdividers. Its cost was but one-fourth that of the Big Sewer.

Peter Warshall studied biology and anthropology in college. He received his Ph.D. in 1973. He co-directed the survey of 400 septic-tank systems and helped write an ordinance for the local Utilities District to care for on-site systems. Since then, he has written papers on waterless toilets and greywater systems as well as consulting on land-retention sewage recycling. He presently is Land Use/Water Editor of *The Co-Evolutionary Quarterly* and co-founder of Warshall-Breedlove Watershed Consultants. His book WATERSHEDS: SOILS, WATER AND LIVING-IN-PLACE will be published by Penguin.